OPERATIONAL DECISION-MAKING IN HIGH-HAZARD ORGANIZATIONS

Operational Decision-making in High-hazard Organizations
Drawing a Line in the Sand

JAN HAYES

Australian National University, Australia

CRC Press
Taylor & Francis Group
Boca Raton London New York

CRC Press is an imprint of the
Taylor & Francis Group, an **informa** business

CRC Press
Taylor & Francis Group
6000 Broken Sound Parkway NW, Suite 300
Boca Raton, FL 33487-2742

First issued in paperback 2017

© 2013 by Jan Hayes
CRC Press is an imprint of Taylor & Francis Group, an Informa business

No claim to original U.S. Government works

Version Date: 20160226

ISBN 13: 978-1-4094-2384-3 (hbk)
ISBN 13: 978-1-138-07477-4 (pbk)

Visit the Taylor & Francis Web site at
http://www.taylorandfrancis.com

and the CRC Press Web site at
http://www.crcpress.com

Contents

List of Figures

List of Tables

List of Stories

List of Abbreviations

ALARP	as low as reasonably practicable
ANSP	air navigation service provider
ATC	air traffic control
ATSB	Australian Transport Safety Bureau
BPR	business process re-engineering
CASA	Civil Aviation Safety Authority
CCPS	Center for Chemical Process Safety
CEGB	Central Electricity Generating Board
CTA	cognitive task analysis
DAP	duly authorised person
GBE	government business enterprise
HRO	high reliability organization
HRT	high reliability theory
HSE	UK Health and Safety Executive
IAEA	International Atomic Energy Agency
ICAO	International Civil Aviation Organization
INES	International Nuclear Event Scale
MATS	Manual of Air Traffic Services
MHF	major hazard facility
MOS	Manual of Standards
NASA	National Aeronautical and Space Administration
NAT	normal accident theory
NDM	naturalistic decision making
OD	Operations Director
OECD	Organization for Economic Co-operation and Development
OEF	operational experience feedback
OHS	occupational health and safety
PDCA	plan-do-check-act
PRA	probabilistic risk assessment
QA	quality assurance
QC	quality control
QRA	quantitative risk assessment
RE	resilience engineering
RPD	recognition-primed decision
SRT	system restoration time
SS	Systems Supervisor
UK	United Kingdom
US	United States of America
WANO	World Association of Nuclear Operators

Foreword

Andrew Hopkins

Emeritus Professor of Sociology, Australian National University, Canberra

Safety science is multidisciplinary, but there are few of us working in this field who can claim to be proficient in more than one of the contributing disciplines. Jan Hayes is someone who can. She is both a qualified and experienced engineer and a qualified and experienced sociologist. This is what makes her work so valuable. She writes here about the thought processes of technical professionals with the sensitivity and clarity that comes from her mastery of these two foundation disciplines.

This book is an enormously valuable addition to the literature on high reliability organizations (HROs). Jan interviewed experienced shift managers – an influential, yet previously little studied group. Her book examines how they go about making decisions, in particular decisions to shut down or modify a process when the level of risk has increased. She shows that where there are applicable rules, they are guided by these rules. In the many situations in which there are no rules they draw on their own long experience. Interestingly, they often formulate this experience as informal rules of thumb about when and how to take action. She shows, too, that they tend not to think in terms of a continuum of risk but rather in more dichotomous terms – safe or unsafe, depending on whether or not all the required hazard control barriers are in place.

The organizations studied in the original HRO research were all identified, making it difficult for the investigators to make critical remarks about them. Perhaps that is why those organizations are portrayed in such glowing terms. In contrast, Jan has tried to provide a more balanced account of the HROs she studied. This was relatively easy in the case of the two anonymous organizations, but the identity of a third organization could not be disguised, and that organization is to be commended for having raised no objection to the publication of the findings.

My own work on major industrial accidents has shown that the path to disaster is paved with poor decision-making. It is refreshing to read a book that examines decision-making processes that protect organizations from disaster.

I was the principal supervisor for the thesis on which this book is based. Jim Reason, of Swiss cheese fame, was one of the examiners. He had this to say: 'I would rate this doctoral thesis as being among the best I have examined during my academic career.'

Preface

This book is based on the view that safety decision-making by operational managers in high-hazard organizations is fundamentally impacted by their experience and judgement. These factors impact directly on the sense that these people make of the situations that develop in their workplaces – in the cases described here a nuclear power station, a chemical plant and an air traffic control operations room. Readers will see that stories provide a way for operational managers to share that knowledge, experience and expertise – with each other and with us in this book.

Of course this 'sensemaking' perspective applies equally to me. In listening to, and making sense of, the stories told by those who participated in the research I have been influenced by my own background and experience. I therefore owe it to readers to explain a little of that background. I worked as a practising engineer for 20 years in a range of high-hazard industries, but mostly offshore oil and gas. From an initial focus on process engineering, I developed an interest in risk management, in particular risk modelling. In the mid-2000s I made a major jump from industry to academia and from engineering to social science, in particular to sociology. The common theme amongst these different career choices is an interest in industrial safety and how accidents can best be prevented.

Stories of accidents have figured strongly in the development of my own professional identity. At the time of the Piper Alpha fire in 1988, I was a young process engineer working in the Australian offshore oil and gas industry and I remember very well the shock of my more experienced colleagues at the news that 167 lives had been lost, and that the entire platform had collapsed into the North Sea. A few years earlier, I had been involved in an incident on an Australian offshore platform in which one of my colleagues was killed. If I needed a further reminder of the hazards of the oil and gas industry, one was available. Ten years after Piper Alpha, a major fire and explosion at the Longford gas plant in Victoria killed two people, injured several and caused so much damage to the facility that natural gas supply was interrupted to millions of domestic and industrial users for several weeks. I had worked at this site as a young engineer and knew personally several of those directly impacted by the fire. What fascinates me in particular is that this industry has so much potential for damage and destruction and yet that potential is effectively controlled and contained almost all the time. This brings me to an interest in how people responsible for operations in this and other high-hazard organizations are generally so effective.

This book is therefore my attempt to make sense of the stories I was told and other organizational experiences I had over weeks at each work site. Although some readers may wish to see it as an attempt to objectify decision-making practices, from a sense-making perspective, it remains my constructed reality, rather than any absolute view. After close to five years of discussing the stories I was privileged to hear with a range of professional people in industries ranging from medicine to offshore oil and gas, I know that the issues I have highlighted have a strong resonance with others.

Academic readers are offered a perspective on decision-making that is both practical and sociological, and I hope that this unusual combination provides some new insights. I invite readers who work in hazardous environments to use these stories of decision-making to reflect on their own experience and how it drives their actions.

Jan Hayes
Melbourne
30 April, 2012

Acknowledgements

Firstly, thanks to Andrew Hopkins. Without his encouragement, this book would not exist. This applies in a specific sense, but also in the sense that his work published over several decades has provided the inspiration for me (and many others) to think differently about the organizational contribution to safety.

Thanks also to Barbara Cox, John Godfrey, Joy Hunter, John Muller and Ron Wilmshurst for feedback and encouragement. And to Valerie Linton and Ankie Larsson of the Energy Pipelines Cooperative Research Centre who have made possible my transition from industry to academia.

Most of all, thanks to the operational managers who generously shared stories of their working lives over the past 25 years and to the organizations that chose to participate in the research and generously allow me free access to their employees, facilities and documentation.

The research described was funded as part of an Australian Research Council Industry Linkage Grant. Approval for the research was received from the Australian National University Faculty of Arts Ethics Committee.

Chapter 1

Introduction

'Safety decision-making by operational people? There's nothing to study because it's all written into our procedures.' This statement from one oil refinery executive is typical of the view held by many managers. In fact, this book seeks to highlight that this is a significant simplification of how people in high performing organizations dealing with complex hazards work to maintain public safety. As we shall see, such organizations rely critically on the professional judgement and expertise of senior operational staff, often without understanding that this is the case. This book seeks to describe how key decisions are made by people working in three high-hazard organizations – a chemical plant, a nuclear power station and an air navigation service provider. In these days of cost cutting and business process management where only things that can be measured are valued, this has important implications for public and worker safety.

Research on decision-making in organizational settings has a long history and has been taken up by a range of academic disciplines including economics, management and cognitive psychology. Decision-making has been widely viewed as a key management process and normative theories have led to the development of a range of practical aids and training courses on improved decision-making. Many of these use the 'classical decision method' involving rational analysis of options in order to make an optimal choice. Since the 1980s, safety decision-making has followed this trend and company systems are now typically based on a specific form of rational analysis that we call risk management. Such methods are apparently so ubiquitous that there may seem to be no need for a book on operational decision-making in particular since there is nothing very interesting to say.

On the contrary, the fundamental proposition of this book is that, despite little understanding or acknowledgement, organizations rely critically on the experience and judgement of professionals such as senior operational staff to keep workers and the general public safe. Risk-based rules and procedures play an important role in ensuring high levels of safety performance but such systems of decision-making replace one large judgement with lots of smaller ones. This book shows that such judgements are critical in the safe operation of complex, hazardous systems and that an excessive focus on process, rather than content, has obscured the role that professional experience and judgement has always played, and will continue to play, in organizational safety.

One group of industrial and infrastructure organizations operates in an environment that makes some decisions especially important. These are high hazard organizations, where the technology or activities involved mean that, if things go wrong, many people could be injured or killed. Air traffic control, nuclear power generation, offshore oil and gas production and petrochemicals manufacturing are all examples of organizational activities where the intrinsic danger is high, even if the risk is low due to the low likelihood of a serious failure. Much safety research in these types of organizations is based on field observations. The literature groans with accounts by social scientists of their experiences in control rooms, flight decks, operating theatres and emergency rooms. Another genre of decision-making research focuses on management decisions made in offices and meeting rooms well removed from the coal face (or control room, flight deck etc). The work described here does not adopt either of these perspectives, but aims to cover the space in between these two views. Our focus is on those decisions made by operational managers. These individuals, who supervise field personnel directly, are the most senior people on shift and provide the link to more senior management remote from day-to-day operations. The following story describes one such individual at work in his job at a nuclear power station.

STORY 1: COULD THE SAME FAULT BE PRESENT IN THE RUNNING REACTOR?

During an outage of one nuclear reactor for routine maintenance and inspection (whilst the other reactor on the site remained operational), an internal weld failure was discovered. This fault was unexpected and the duty shift was heavily involved with engineering and maintenance personnel in deciding how the repairs were to be carried out on the offline reactor.

Shift Manager Interviewee 4 came on duty at a later stage and asked the technical specialists two questions: could the same fault be present in the running reactor; if so, what is the worst credible damage that could be present? The specialists could not be sure that the running reactor did not have the same fault and felt that they could not discount the possibility of a significant gas release as a result (although there was no question of loss of radiation containment).

Based on the advice he received, Interviewee 4 decided to shut down the running reactor. This reactor was found subsequently to have a similar fault, but engineering analysis showed that there was no potential for gas leakage.

Interviewee 4 describes the outcome: 'In a nutshell if you just want to look at it on the balance sheet, I took the reactor off for no good reason ... and we were off for a week, so I cost the company a lot of money there. Do you see what I mean? In hindsight that was a week's worth of generation lost because I decided to come off because I had this gut feeling that I wasn't happy with it ... he [the site manager] openly congratulated me on making the right decision.'

This single story illustrates several key issues. In this case, the operational manager made a very conservative choice. The traditional cautionary tale in the safety literature would have a different ending with the overall conclusion that he averted a near-disaster by his actions. In this case, the actual result was more banal and yet this story is perhaps more interesting because it does not take the expected line. Nuclear power station operations are highly proceduralized, but in identifying the potential problem and choosing this course of action, this operational manager had no procedure to follow. His actions were a direct result of his professional and organizational environment. We will explore the factors behind decisions such as these – the perspective on system operations that led to this operational manager seeing what he thought was the potential for an accident, his confidence in interrupting operations based on that potential and the response of the organization to the outcome. We will also consider what constitutes a good decision. In more everyday circumstances, we tend to judge decisions by their outcome and yet by that measure the operational manager in the story above made a poor choice. In seeking excellent safety performance, different measures are needed to judge the quality of actions taken.

Before looking further at decision-making practices, there are two further introductory issues to be addressed. The first is why we should be interested in looking at decision-making in cases where there has been no major problem. People in industry love hearing stories about disasters (provided that they happen to other people and the focus is on technical, rather than personal, details). Putting aside the question of *schadenfreude* (or taking pleasure in the misfortune of others), such examples can provide important, concrete lessons about what not to do. This book could have focused, for example, on problems with operational decision-making in the cases of two recent blowouts in the offshore oil and gas industry: Deepwater Horizon (Hopkins 2012) and Montara (Hayes 2012). Instead, in the tradition of high reliability research, the focus is on lessons that can be drawn from cases where things go right.

High Reliability Theory (HRT) attempts to explain how successful high hazard organizations manage to operate in a way that is generally failure free. Rather than focusing on accident analysis, theorists in this field look at how organizations behave in order to minimise both the number and severity of incidents. HRT focuses on the organizational qualities that are required to achieve 'mindfulness', which is seen as the key to a high level of safety performance (Weick and Sutcliffe 2001). Organizations that achieve this are known in this field as High Reliability Organizations or HROs. This book is in that tradition of 'normal operations studies' (Bourrier 2002, 2011). More detail on high reliability theory and its relevance to operational decision-making can be found in Chapter 2.

The final introductory point is to highlight the role played by stories in this book. This case study based research generally follows ethnographic methods of the kind widely used in sociological research (Silverman 2001), in sensemaking research (Weick 1995) and in safety research (Hopkins 2006). Most data collection was via semi-structured interviews, although workplace observations and document review also contributed. Interviews were recorded, transcribed and reported in the form of a series of stories detailing the experiences of operational managers with extensive direct quotations from the people involved. Reporting data in the form of stories was a deliberate choice. Dreyfus's (1986) model of expertise and learning suggests that, whilst universal rules and generalized models are very important when one starts to learn a new skill, once an individual acquires some experience in any given field, his/her efforts move to using context-specific information to determine how best to achieve the desired outcome. Actions are based on experience and intuition, not conscious application of logic or rules. Flyvbjerg (2001) points out that case study research can provide a rich source of input to our development of context-specific experience and intuition – our mental models. Case study research is therefore an important way of increasing the store of human knowledge. The book aims to explore the individual and organizational context of safety decision-making in situations where production pressures are also ever-present; and to present the results of the work in a case study form that makes it easy for other experienced safety practitioners to add to their own understanding of these issues.

It is perhaps worth acknowledging at this point that this type of knowledge may be unfamiliar to some readers. In the physical sciences and in engineering, much use is made of relevant theories, and generally these can be expressed quantitatively, often in mathematical terms using equations. The situation in the social sciences is quite different in that 'theory' can also refer to a body of knowledge in qualitative terms. More detail on learning, stories and case studies can be found in Chapter 2.

This book is structured into 11 chapters.

Chapter 2 gives some theoretical background on the key ideas used in the analysis of decision-making.

The detailed content of the book is divided into two parts. In Part A, Chapters 3, 4 and 5 record the stories told by operational decision makers working in a nuclear power station, a chemical plant and an air navigation service provider.

Part B of the text examines the details and context of those stories. Chapter 6 discusses the dual organizational identities of the operational managers – as employees and as professionals. Chapter 7 addresses how rules are used by this group. Chapter 8 looks further at professionalism and how it impacts decision-making. The impact of relationships with peers, subordinates and managers is discussed in Chapter 9. Chapter 10 looks at the form of the experienced-based judgements made by the operational managers.

Chapter 11 includes a summary of the key arguments in the book and a discussion of the implications for organizations and for regulators, including some suggestions of how to support operational decision-makers in practical ways.

Chapter 2
Theoretical Perspectives on Making Safe Decisions

This chapter summarizes the key theoretical perspectives that form the basis of the analysis of decision-making practices in the remainder of the book. The aim is to provide background for those less familiar with these ideas and/or those who are particularly interested in the theoretical foundations of the work. The analysis in this book draws on each of these in making sense of the stories told by the operational managers.

As described in the Introduction, the main theoretical context to this research is High Reliability Theory (HRT). Work in this field has highlighted several key qualities of high performing organizations that are relevant to decision-making and these are discussed below in Section 2.1. Section 2.2 briefly describes the rather similar approach taken by Resilience Engineering (RE) researchers and the implications for operational decision-making. In contrast, many companies use the language of risk management to describe their decision-making and this is also discussed, within the context of classical decision theory, in Section 2.3. Theories of accident causation are another important aspect of safety research and what they have to say about decision-making is summarized in Section 2.4. The most relevant is James Reason's Swiss cheese model.

Our decisions are based on what we know and hence are intertwined with how we learn. Theoretical perspectives on how experts learn are described in Section 2.5. Finally, the perspective of sensemaking is reviewed in Section 2.6. This way of thinking about decisions emphasizes the role of past experience, not from the perspective of rational deliberation but as a way of describing what experienced people notice in complex situations and how that translates to action.

2.1 High Reliability Theory

This body of work has its origins in analysis of the 1979 Three Mile Island nuclear power station incident in the US. The original ideas (La Porte 1981) were published in a set of essays (Sills et al. 1982) that also includes Perrow's early work on Normal Accident Theory, which is described later in this chapter. Researchers studied organizations that are required to operate at very high levels of safety. La Porte (1991) also acknowledged a second operational challenge of the group of organizations defined as High Reliability Organizations or HROs: 'to maintain the capacity for meeting intermittent, somewhat unpredictable, periods of very high peak demand and production'. (La Porte 1996: 60) Rochlin (1993) reviewed the participating organizations and established a set of six criteria that provides a working definition of an HRO:

1. The organization is required to maintain a high level of safety performance if it is to be allowed to continue to operate.
2. The organization must also maintain high levels of capability, performance and service to meet public and/or economic expectations and requirements.
3. Because of the consequences of error or failure, the organization cannot easily make marginal trade-offs between capacity and safety. Safety is not fungible.[1]
4. As a result, primary task-related learning cannot proceed by trial and error since the first error may be the last trial.
5. The technology and primary task are both so complex that safety and capacity issues must be actively and dynamically managed.
6. The organization will be judged to have failed and will be criticized almost immediately if either the safety performance or service/product delivery degrades.

More recently Karlene Roberts (another member of the original Berkeley team who still works in this field) described an HRO as being 'an organization in which errors can have catastrophic outcomes, but which conducts relatively error free operations over a long period of time making consistently good decisions resulting in high quality and reliable operations'. (Bourrier 2005: 94) It is argued by high reliability researchers that organizations such as nuclear power stations already operate at extraordinary levels of safety performance (La Porte 1996). The aim of their research is to identify those facets of organizations that lead to this high reliability, that is, a demonstrated greatly reduced potential for serious accidents.

Early work on high reliability organizations (La Porte and Consolini 1991) focused on decision-making as one of three areas where organizations with the potential for catastrophic failure were likely to differ from other (low reliability) organizations. (The other areas were structural responses to hazards and peak loads, and the tightly coupled, interdependent nature of operations.) LaPorte et al. (1991) describe the challenges of decision-making for HROs as:

- Extending the rational decision-making process as far as possible within the constraints of the data and time available for operational decision-making and mandatory adherence to formal documented operating procedures.
- Being sensitive to areas where incremental decision-making based on judgement must be used.

---

1   Fungible is a legal/contractual term meaning that which can be exchanged for like goods.

- Being alert for small errors or unexpected events that have the potential to escalate into catastrophic failures.

The challenge of identifying those decisions that have the potential for catastrophic consequences remains. These issues were investigated by the HRO Project at Berkeley that developed a number of findings about decision-making in HROs based on field observation and interview (La Porte 1996), specifically:

- The organizational structure is typically hierarchical during normal operations, but as the tempo of operations increases, structures become more collegial and based on expertise. Decision dynamics also become more fluid as tempo rises.
- Decision-making tends to be decentralized to the level where actions must be taken.
- Once made, decisions are quickly implemented.
- HROs 'exhibit a quite unusual willingness to reward the discovery and reporting of error' (La Porte 1996: 64) and are always looking for ways to improve.

HRO research has been undertaken (primarily in the US) in a wide variety of high hazard industries and facilities such as aircraft carriers (Roberts 1990, La Porte and Consolini 1991, Weick and Roberts 1993, Roberts et al. 1994), nuclear power stations (Schulman 1993, Carroll 1998, Carroll et al. 1998, Carroll et al. 2001), chemical plants (Carroll 1998, Carroll et al. 1998), air traffic control systems (La Porte and Consolini 1991), offshore oil and gas platforms (Flin 1996) and electricity distribution (Roe and Schulman 2008). By 1999, a large body of field research had been produced with the general aim of investigating how HROs generally manage their operations to achieve a safe outcome. The findings of this body of work were synthesized into a single theory (Weick et al. 1999, Weick and Sutcliffe 2001) describing HROs as organizations that exhibit mindfulness, a state of organizational learning which fosters the capability to discover and manage unexpected events. Mindfulness is characterized by the following five qualities:

1. Preoccupation with failure,
2. Reluctance to simplify interpretations,
3. Sensitivity to operations,
4. Commitment to resilience, and
5. Deference to expertise.

Each of these qualities is described further overleaf.

Preoccupation with Failure

If failure is a precondition to organizational learning (Sitkin 1992), HROs have limited opportunities for learning, since failure in a broad sense is to be avoided at all cost. As Carroll (2002) points out, the challenge for HROs is to avoid both the catastrophes associated with trial and error learning and the complacency that can arise when attempting to learn from successes. Effective HROs are preoccupied with failure in that they see thorough analysis of any and all small failures as essential for learning (Sitkin 1992, Weick et al. 1999, Weick and Sutcliffe 2001). Small local failures are seen as indicators of overall system health. Since an important part of the role of maintenance is to deal with failures (at least those related to equipment), maintenance departments in effective HROs have a much more central role than in traditional organizations. The experience of maintenance personnel is valued for the breadth of understanding of organizational strengths and weaknesses that it brings (Bourrier 1996). Incident investigation is valued for the same reason (Carroll et al. 2001).

Reluctance to Simplify Interpretations

Perrow (1999: 9) makes the point that 'seeing is not necessarily believing: sometimes, we must believe before we can see'. Effective HROs are aware of this and seek a diversity of views on safety issues. Schulman (1993) describes what he sees as 'conceptual slack' in effective HROs where an organization seeks to have a variety of analytical perspectives regarding assumptions, theories and models of the technology or production processes in use. Based on his research at the Diablo Canyon nuclear power plant, he asserts that reliability is achieved, not by organizational invariance and the strict following of fixed and formal rules and procedures, but by actively maintaining multiple views (from various departments) over site activities. This is not to say that the nuclear power plant has no procedures – in fact it has hundreds of them – but they are regularly reviewed, updated and changed. In addition, the organizational culture supports and protects the necessary organizational ambiguity required to deal with uncertainties in the assumptions that underlie the formal systems.

Sensitivity to Operations

Sensitivity to operations in HROs is the formal, theoretical term for the more practical expressions 'having the bubble' from the US Navy and 'situational awareness' from pilots and air traffic controllers. It means that one or more individuals have an overall understanding of the state of the operational system. They also have sufficient expertise to see patterns in small anomalies that arise, so that problems can be anticipated before they develop. This is essentially the quality that Perrow (1999) claims is missing in highly complex technologies, where unexpected failures and unexpected interrelationships mean that no-one within the organization has a clear understanding of cause and effect in operations. This forms part of the basis for his Normal Accident Theory described below.

Commitment to Resilience

In engineering terms, some facets of resilience are covered by the defence in depth concept – that the potential for accidents with complex causal chains is addressed most effectively by a suite of defences aimed at elimination (of potential causes of hazard), prevention (of incident occurrence), mitigation (of the effects of an incident) and recovery (from the effects of an incident). For HROs, resilience covers not only these planned layers of defence against accidents, but also the ability to be able to respond in the moment to unexpected events. (Wildavsky 1988) defines resilience as the ability to bounce back. He promotes development of active resilience, that is, 'a deliberate effort to become better at coping with surprise'.

This aspect of HRT has become central to the work of the Resilience Engineering researchers described in Section 2.2.

Deference to Expertise

Weick (1999) found that successful HROs typically have a formal hierarchical structure but that organizational processes are flexible and allow responsibility for decision-making under high pressure situations to move to experts who are close to the field of action. Turner (1997) also proposed that the orderliness that organizations seek is a two-edged sword. Whilst order generally makes processes and outcomes more stable, predictable and consistent, it can also amplify the small errors that will always occur and disseminate them through the system, with unpredictable consequences. The spread of what Turner calls anti-tasks can be greater in organizations that are better organized. Weick (1998) suggests that this might be why successful HROs seem to exhibit some of the disorganized characteristics of the garbage can model (Cohen et al. 1972). In the garbage can model, choices, problems, solutions and decision makers move independently through the system and come together based only on their concurrent presence within the system. The arrival timing of elements into the system, and therefore possible linkages, can be influenced by organizational structure but problems and solutions are not linked by intention based on cause and effect. This model is generally associated with poor organization, but perhaps some degree of randomness interrupts the propagation of anti-tasks through the system.

Weick and his fellow researchers show that attainment of the five qualities discussed above will lead to a state of organizational mindfulness, which in turn fosters the 'capability to discover and manage unexpected events'. Such a capability will lead to reliability. In HRO theory, reliability is the capacity to repeatedly produce a particular and specific outcome (in this case, no accidents). It is thought that this is produced not by stable and unvarying activity, but 'what seems to happen in HROs is that there is variation in activity, but there is stability in the cognitive processes that make sense of that activity'. (Weick et al. 1999: 87) This is in marked contrast to the concept of quality management (see Chapter 7), which is based on the premise that the key to uniform outcomes is uniform activity.

Other high reliability researchers have tentatively accepted the five-process model but there is some evidence that organizations achieve high reliability by a variety of methods. Nuclear power generation in particular has been found to adopt different techniques in a number of areas. Schulman (1996) found cases in some hazardous industries such as air traffic control, chemical plants and fossil fuel electricity utilities, where organizational heroes were the subject of tales told to reinforce and transmit organizational values and goals related to problem solving. In contrast, his research at nuclear power plants uncovered an attitude that he describes as anti-heroic, where bull headed individuals who are likely to take independent action in the event of an emergency are not welcome. He proposes a two-parameter model to explain this behaviour, citing nuclear power as an industry where the level of analysis and the level of action required for recovery from abnormal operation are both system-wide and hence must be enacted in an integrated fashion, not by an individual hero. In contrast, some HROs (or some situations, in some HROs) require only local analysis and local action to recover from abnormal operations. In Schulman's view, one example is air traffic control, where even a system wide failure (such as loss of standard voice communications) requires detailed analysis and action by the air traffic controllers responsible for traffic in each individual sector that is at a local level. Bourrier (1996) has also highlighted some differences between the strategies used in nuclear power plants in the US and those in France to achieve high reliability.

The five-parameter model represents a major shift in HRO research. The rich body of case studies has been distilled into a general descriptive model and the model has expressed the results of the research in a way that can be seen not only as descriptive, but also as predictive. Other organizations that seek to improve their performance generally, not just in the area of safety, are encouraged to pursue the five identified qualities. Weick et al. conclude: 'In a dynamic, unknowable, unpredictable world one might presume that organizing in a manner analogous to HROs would be in the best interest of most organizations.' (Weick et al. 1999: 87) Far from being applicable only to a small group of exotic organizations, high reliability research is portrayed as providing lessons for any organization seeking to operate in difficult conditions.

So how does high reliability theory help us to understand operational decision-making? This large body of research highlights a range of organizational and individual attitudes that promote excellent safety performance and, as we will see, some of these are seen in the organizations studied and are relevant to operational decision making practices. In describing a general model of high reliability, Weick et al. have emphasized the cognitive processes required to meet organizational safety goals (such as no accidents) in complex circumstances. This model applies equally to production goals so, for example, an organization will have a more reliable performance in meeting production targets or operational goals if it is preoccupied with the possibility of failure to meet those goals. This conceptualization of HRT, however, is silent on the issue of conflicting goals which was part of the original body of high reliability research (see La Porte 1996 quoted above). This book does not seek to build on this model or theory directly, but rather to provide three new case studies that focus attention again on conflicting goals in successful high hazard organizations and how they are managed.

2.2 Resilience Engineering

The new century has seen the development of another new body of safety theory known as Resilience Engineering (RE). Hollnagel defines resilience as 'the intrinsic ability of an organization (system) to maintain or regain a dynamically stable state, which allows it to continue operations after a major mishap and/or in the presence of a continuous stress'. (Hollnagel 2006: 16) Conceptualizing safety as a search for resilience also has much in common with Wildavsky's (1988) work *Search for Safety,* in which resilience (the capacity to cope with unexpected events – the ability to bounce back) is contrasted with anticipation (the effort made to predict and prevent potential disasters before damage is done). Today's resilience engineering theorists – see for example (Hollnagel et al. 2006, Hollnagel et al. 2008, Hollnagel et al. 2011) – emphasize the need for high performing organizations to have both the ability to anticipate and plan, and also the ability to adapt and respond.

It could be argued (as Hale (2006) has done) that this work has much in common with the thinking of the high reliability theorists described in Section 2.1. Despite the familiar starting point RE researchers have re-invigorated the search for decision-making strategies that assist organizations to balance conflicting goals. Woods describes the organizational trade-off between production and safety as a 'sacrifice decision' and comes to the conclusion: 'Resilience Engineering needs to provide organizations with help on how to decide when to relax production pressure to reduce risk'. (Woods 2006: 32) Another aspect of RE that is relevant to a study of operational decision-making practices is Hollnagel's view that 'both failures and successes are the outcome of normal system variability'. (2008: xii)

These and other concepts from theorists in the RE school are further referenced in later chapters.

2.3 Classical Decision-making and Risk Management

A very different body of knowledge about decision-making in general is classical decision-making or rational choice theory. This is relevant to decision-making practices by operational managers in high hazard organizations because the risk management processes that are widely used in industry are based on the concept of rational choice theory.

2.3.1 Rational Choice Theory

The traditional theoretical approach to decision-making as described in much of the economics and management literature takes a cognitive approach, dividing all decision-making into four generic steps as described by Flin (1996):

1. Identify the problem,
2. Generate a set of choices or possible solutions,
3. Evaluate each option (using a wide range of strategies), and then
4. Select and implement the best option.

This model is the cornerstone of decades of decision-making research and is generally known as rational choice theory. Decision-makers are seen as rational actors – individuals who make choices based on logical analysis of available options. Much of the research has focused on steps three and four of the generic model that is seeking the most appropriate way to evaluate the available options and hence determining which should be defined as best. A common example is to identify a set of criteria that represent all the different features of the listed options and to develop a weighting for each criterion. Each option is then rated against each criterion and total scores developed based on the cumulative total of rating multiplied by weighting for each criterion and for each option. The best option is thus selected. Rational choice theory has been so influential that some researchers have even postulated that it is capable of forming the unifying discipline for the whole of social science (Schram and Caterino 2006).

Despite the enthusiasm of some researchers, others have identified many situations in which people follow classical decision theory in only the most approximate fashion. Researchers in the traditional mode of decision-making see the four point model described above as normative. Any observed behaviour that varies from this model in any particular case is due to the cognitive complexity of the evaluation required. If there are several options available to the decision maker and each has a range of advantages and disadvantages, then the mental assessment required to determine the best option using the rational choice method quickly exceeds our capacity. The idea that we have insufficient cognitive capacity to assemble and evaluate all the necessary facts in the case of complex decisions is known as 'bounded rationality'. (Simon 1956) Research efforts have been directed at the identification of rules of thumb that individuals develop to take shortcuts in the process. Whilst such rules may be cognitively economical, they are seen as biases, and the price of their use is seen as increased potential for error. In an attempt to counteract these effects, further research efforts have been directed into development of training methods to assist individuals to make decisions by mental processes that more closely resemble the rational choice method. A well-known example is the Kepner-Tregoe Matrix which has its origins in the Rand Corporation and has been used in manufacturing and other management circles since the 1960s (see Kepner and Tregoe 1997).

Researchers in this field see the advantage of the rational choice model as generating a result based solely on logic and analysis. The aim is to remove judgement and intuition from the process. Critics of this model (Turner 1990, Reed 1991, Tetlock 1992, Carroll 1993, Klein 1998, 2003, Argyris 2004) point out that, in reality, the decision-maker is required to make many small decisions based on judgement and intuition in order to generate a range of options to consider, the list of evaluation criteria, their relative weightings and the scores of the individual options. A single judgement on a large scale has been replaced by many judgements on a smaller scale for the same choice to be made.

The other major limitation of this method of decision-making is that it assumes significant time is available to the decision-maker to identify several options and consider the pros and cons of each one. Clearly, in an operational situation actions are often time pressured and time for deliberation may be very short. This limits the usefulness of rational choice approaches in operational situations and their validity as a preferred decision-making approach. Nevertheless, rational choice in the form of risk management is widely applied to safety decisions.

2.3.2 Risk management – rational choice applied to safety

The ultimate application of rational choice theory to safety decision-making in high hazard industries is the use of Quantitative Risk Assessment (QRA) and the associated decision-making principle 'as low as reasonably practicable' (ALARP). In the UK nuclear industry, this is called PRA (Probabilistic Risk Assessment), but the techniques and principles are essentially the same.

QRA is a mathematical technique that originated in the nuclear industry in the 1970s. Since that time, its use has spread to other high hazard industries such as offshore oil and gas, onshore chemical and petrochemical facilities, transportation of hazardous goods and aviation. The technique attempts to determine a numerical estimate of the frequency of fatality (of either workers or the public or both) associated with the facility, operation or activity in question. This is typically done by development of all possible causal chains that could lead to fatality and estimation of both the potential consequences (size of fire, extent of structural damage and so on) and the likely frequency or probability of each step in the causal chain.

When faced with a choice, the results of a QRA can assist in decision-making by providing a numerical ranking of the options based on the estimated frequency of fatality (a measure of risk). This analysis itself is subject to a range of uncertainties in representing complex issues by a single index. Even once these difficulties have been overcome, two significant questions remain unanswered:

- How is the cost of each option taken into account? In other words, if the lowest risk option is more expensive to adopt than other options, which should be chosen?
- Is the absolute risk acceptable? In other words, does the lowest risk option available still introduce an estimated frequency of fatality that is too high?

The process used to address these questions is comparison of the calculated risk with fixed criteria followed by cost benefit analysis. Specifically,

- There is a level of risk to an individual that is deemed to be intolerable. If the risk is found to be above this value, changes must be made in order to reduce risk, regardless of cost, except in the most exceptional of circumstances.
- There is a lower (but non-zero) level of risk that is deemed to be broadly acceptable. At this risk level (and below), risk should be monitored to ensure that no significant increase occurs, but further expenditure on risk reduction is not justified.
- Between these two risk levels is what is known as the ALARP region. If risk falls into this region, it should be reduced to a level that is as low as reasonably practicable. Risk reduction measures must be identified and evaluated in terms of cost (money, time and effort) and possible risk benefit. Measures should be put in place provided that the cost is reasonable when compared to the benefit gained.

The comparison of pros and cons often takes the form of a numerical cost benefit analysis where a notional financial value is assigned to each cost and benefit contributor, including a numerical value for notional value of statistical lives saved.

The origins of the concept that organizations have an obligation to reduce risk to a level that is as low as reasonably practicable are in law, rather than management or engineering, and this idea has been widely adopted in safety legislation. There is no simple definition of the legal expression 'reasonably practicable' but legal interpretations in Australia and the UK can be traced to the definition by Asquith LJ in *Edwards v National Coal Board (1949)* quoted in Bluff and Johnstone (2004: 8):

'Reasonably practicable' is a narrower term than 'physically possible' and seems to me to imply that a computation must be made by the owner, in which the quantum of risk is placed on one scale and the sacrifice involved in the measures necessary for averting the risk (whether in money, time or trouble) is placed in the other; and if it be shown that there is a gross disproportion between them – the risk being insignificant in relation to the sacrifice – the defendants discharge the onus on them. Moreover, this compensation falls to be made by the owner at the point of time anterior to the accident.[2]

Not all risk management processes rely on generation of numerical data for judgements about ALARP. Processes and procedures for decision-making based on qualitative risk management processes usually also follow classical decision-making concepts, where cause and effect relationships are assumed to be knowable and available for use as the basis by which the best choice can be made, even if no numerical calculations are involved. Such risk management processes are subject to the same criticisms as those levelled at the rational approach generally and described earlier. As Hopkins points out, techniques such as cost benefit analysis 'do not generate a set of rules which can be automatically applied to determine whether the risk is as low as reasonably practicable and ultimately a judgement must be made in each situation about what a reasonable employer would do'. (Hopkins 2005b: 113) In a very different context, in their review of recent updates to US Environment Protection Agency technical risk assessment protocols, Abt et al. (2010) have emphasized the importance of understanding that risk assessment is a method of evaluating public policy options, but not a means of selecting policy options. In other words, risk assessment is 'a means to an end rather than an end in itself'. Similarly, the nuclear industry would seem to be aware of the strengths and possible misapplication of risk assessment. An industry procedure on decision-making warns that 'risk assessment tools provide input for managers to evaluate decision options, but do not replace management judgement'. (World Association of Nuclear Operators 2002)

From the perspective of safety, risk can be defined as 'the possibility of danger'. (HSE 2001) Most risk management systems assume that consideration of risk must be based on rational analysis of the causes and their effects, but this is not inherent in the definition. Some theorists have expanded the concept of risk to include views of stakeholders in addition to (supposedly) objective, formal and causal analysis – see for example Sandman (1993). When making large investment decisions, such as justification for a new major project, some organizations now consider the rationality of the project as viewed from a range of perspectives other than just their own. This could be seen as acknowledgement of the contextual nature of rationality, but in reality it is usually undertaken as a method of managing secondary risks such as reputation. As Power states, 'the risk management of everything is characterized by the growth of risk management strategies that displace valuable – but vulnerable – professional judgement in favour of defendable process'. (Power 2004: 10)

Power has also expressed serious reservations about the overuse of risk assessment. In his view, 'the experts who are being made increasingly accountable for what they do are now becoming more preoccupied with managing their own

---

2   Unfortunately this judgment really does advocate consideration of risk 'ON one scale' and sacrifice 'IN the other' (my emphasis).

risks. Specifically, secondary risks to their reputation are becoming as significant as the primary risks for which experts have knowledge and training. This trend is resulting in a dangerous flight from judgement and a culture of defensiveness that create their own risks for organizations in preparing for, and responding to, a future they cannot know.' (Power 2004: 14)

Classical decision-making in the form of risk assessment provides an important regulatory and organizational context for the decisions under study here. It also provides a rational decision-making framework for balancing competing goals and for addressing non-operational activities. The key element missing from this perspective is that it does not recognize the judgement required to convert the results of analysis into action. This perspective also assumes that there are no constraints on the time available to decide on a course of action. For operational managers, confronted by complex, dynamic situations in the field, time may be a luxury that they do not have.

2.4 Theories of Accident Causation

Many researchers have turned to analysis of past accidents in an attempt to understand how best to prevent accidents for occurring in the future. Readers may be familiar with some of these safety theories that are based on understanding accident causation. The four most influential are discussed here, along with their implications for decision-making by operational managers.

2.4.1 Normal Accident Theory

An influential theory on the origins of accidents in high hazard industries is Perrow's Normal Accident Theory (NAT). Perrow studied the causes of the 1979 Three Mile Island incident in the US. His initial work was published in a 1982 analysis of the human factors associated with this specific incident (Sills et al. 1982). Perrow (1999) later expanded this analysis to a general theory which argues that some technologies are both so complex and so tightly coupled that accidents are inevitable, to be expected and hence should be seen as 'normal' in those industries. Complex technologies lead to unexpected interactions between seemingly independent parts of the system, and the tight coupling of the system leads to rapid escalation before diagnosis and intervention is possible. Perrow's assessment of which technologies are most vulnerable to normal accidents is independent of the skills, experience, management processes or behaviours of the organization managing the technology, or of the individuals involved. The only variable is the technology itself. NAT has been criticized for its fatalistic and ultimately unhelpful input to organizations that use complex technologies (Roberts 1990, Hopkins 1999, 2001, Dekker 2004). The only conclusion possible from Perrow's work, and a conclusion that he indeed reaches, is that some technologies such as nuclear power generation are simply too dangerous for society to use.

Normal Accident Theory provides little guidance for operational decision-makers in high hazard industries once strategists and policy makers have allowed the industry to exist.

2.4.2 Disaster Incubation Theory

Another theory of accident causation is the disaster incubation theory of Turner (1997) originally published in 1978. In Turner's view, accidents can best be understood as 'cultural disruption'.

Turner's theory (1997) is similar to Reason's Swiss cheese model (see p. 18) in that he postulates that there are common themes across disasters in the period leading up to the disaster (not just in the response). He proposes that disasters are caused by the failure of foresight. He sees that errors accumulate due to a range of communication errors (information known but ignored or distrusted, information buried or distributed across the organization and hence not assembled). The theory is fundamentally optimistic (in a way that NAT is not) as the key to disaster causation, and hence prevention, lies with people and organizations and is not inherent in the technology. Similar to work based on the Swiss cheese model, analysis of past accidents from the perspective of disaster incubation theory has led to lists of generic features and factors that are accident precursors. Minimizing these is the assumed accident prevention strategy. Turner's view was radical in the 1970s and was largely ignored. Despite its interesting perspective on accident causation and the origins and propagation of organizational error, it does not directly illuminate issues of operational decision-making.

2.4.3 Error Management

The error management approach to safety is based on the assumption that a better understanding of the causes of human error will lead to strategies to reduce and contain errors, which will in turn lead to better safety performance. This perspective focuses on individuals and their immediate tasks and workplace environment.

Rasmussen (1982) has studied decision-making in a control room setting and proposed a three-tier model where decisions are characterized as skill-based (almost automatic), rule-based (following a rule or procedure) or knowledge-based (creative). This work has been extremely useful in identification of the appropriate strategy to address the potential for human error in the field in a range of situations. Equipment design, layout, task design and training are all possible error management strategies (Reason 1990). The expanding use of computer applications spawned the sub-discipline of usability and human-machine interface (HMI) design which also commonly uses an error management approach. Whilst this body of work has important safety implications, it addresses those cases where the task itself is well defined and the outcome of a particular action can be judged to be successful or in error unambiguously and within a short timeframe. In fact one aspect of error management is design of systems for good error detection to ensure that this is the case. Many operational decisions are much more ambiguous than this and hence classifications of this type are not helpful.

2.4.4 The Swiss Cheese Model

Organizational psychologist James Reason has written widely about his 'Swiss cheese' model of organizational accidents (Reason 1997, 2008). This model is widely used in industry to facilitate thinking about accident investigation and prevention.

In this way of thinking about accidents (illustrated in Figure 2.1), there is a range of defences in place that are functionally designed to prevent any given hazard from leading to a loss of some kind (such as an accident). In practice, these defences are imperfect (like holes in Swiss cheese). The various hardware and procedural measures in place ensure that failure of any individual measure is not catastrophic. An accident occurs when the holes in the cheese line up and provide an accident trajectory through all the defences. In this model the holes in the cheese have two interesting features. Firstly, they may be due to active failures, for example an operating error that leads to a lower temperature (as at Longford (Hopkins 2000)) or higher level (as at Texas City refinery (Hopkins 2008)) than normal in part of the plant. Alternatively, the holes may represent latent failures. Latent failures are weaknesses in the system that do not, of themselves, initiate an accident, but they fail to prevent an accident when an active failure calls them into play on a given day. Problems arise when such failures in the system accumulate – maintenance is not done, records are not kept, audits are not done. The consequence of a small active failure can then be catastrophic as the protective systems fail to function as expected.

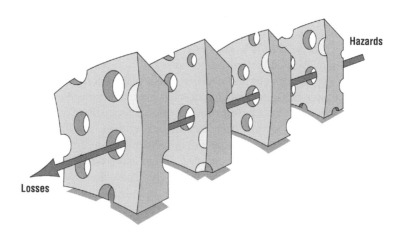

Figure 2.1   Swiss cheese model

Source: From (Hopkins 2012) Reproduced with permission

The second quality of the holes in the Swiss cheese is that they are a function of the organization itself. In this model of accident causation, operator actions in the field are linked to workplace factors such as competency, rostering, control room design, task design and so on, and these issues are linked to organizational factors such as budgets, safety priorities and management styles. In this way of thinking about safety defences, the performance of all components in the system is interlinked.

The benefit of this model to practical decision-makers is perhaps its emphasis on each person as part of the organization as a whole. Whereas classical decision theory sees decision-makers as free agents able to choose any option, the Reason model moves the focus to the organization where the behaviour of individuals is dependent on interactions with others. The model also shows that the full consequences of decisions made may have a long incubation period and that decision-makers may be held accountable long into the future. This was emphasized in the legal response to the Gretley coal mine disaster (Hopkins 2005a). In this case, not only the mine manager at the time of the disaster, but also his predecessor, were found guilty of failing to exercise due diligence. Attempts to use the model in a directly predictive manner (see for example Reason's work on General Failure Types (1997)) have been less successful, certainly in terms of general industry uptake.

The work described in this book is generally consistent with the views on safety and accident prevention put forward by these theorists, however none of the four models of accident causation described in this section assists directly in understanding decision-making by operational managers. The most relevant is the Swiss cheese model as it emphasizes that individuals always make decisions within an organizational context, but it provides no specific insights into how particular decisions are made.

2.5 Experts and Learning

There are many theories linking expertise and cognition. Of particular relevance to operational decision-making is an understanding of how expertise links to use of rules and to learning as described in the work on expertise by Dreyfus and Dreyfus (1986). They describe five stages of skill acquisition for decision-making related to unstructured problems, that is those where there is a potentially unlimited set of relevant information and actions which interact to determine outcomes in ways that are perhaps unclear. The five stages are as follows:

1. Novice
   A novice recognizes relevant facts in a context-free manner. A novice follows rules that are absolute (not context specific).

2. Advanced beginner

   An advanced beginner follows rules, some of which are context specific, based on their limited experience in application of rules. They feel little responsibility for the outcome of their actions, only in their competence in following rules.

3. Competent

   A competent person has a specific goal in mind and selects context-specific information that seems to be relevant. They use rules to determine how best to tackle the problem and select the best course of action. Whilst the course of action is chosen rationally (that is by analysis and application of rules) the decision-maker now feels responsible for outcome.

   This is the highest level of rational problem solving. Dreyfus and Dreyfus call this the Hamlet model; 'the detached, deliberative, and sometimes agonizing selection among alternatives'. (Dreyfus and Dreyfus 1986: 28)

4. Proficiency

   At this skill level, salient features of the task are selected and organized intuitively, that is based on experience. Action selection is deliberate, conscious and based on analysis. Problem recognition is intuitive at this level.

5. Expert

   For an expert, both problem recognition and action selection are intuitive. An expert just acts. They are unaware of their skill – they simply 'drive', or 'fly' rather than driving the car or flying the plane.

A summary of the five stages of skill acquisition is shown in Table 2.1 below.

Table 2.1   Five stages of skill acquisition (Dreyfus and Dreyfus 1986: 50)

Skill Level	Components	Perspective	Decision	Commitment
Novice	Context free	None	Analytical	Detached
Advanced Beginner	Context free and situational	None	Analytical	Detached
Competent	Context free and situational	Chosen	Analytical	Detached understanding and deciding. Involved in outcome
Proficient	Context free and situational	Experienced	Analytical	Involved understanding. Detached deciding
Expert	Context free and situational	Experienced	Intuitive	Involved

The five stages represent a 'progression *from* analytic behaviour of a detached subject, consciously decomposing his environment into recognizable elements, and following abstract rules, *to* involved skilled behaviour based on holistic pairing of new situations with associated responses produced by successful experiences in similar situations'. (Dreyfus and Dreyfus 1986: 35 emphasis in the original) This evolution from the abstract to the concrete reverses childhood learning sequences that require understanding of concrete examples before progressing to abstract reasoning. It also emphasizes the importance of context-independent rules dictating a specific course of action in skill acquisition and decision-making. According to this research, such rules are critical in the initial stages of learning. As experience grows, the external form of such rules becomes less and less relevant. An expert has built an intuitive and context-specific picture. Whilst the picture had its origins, perhaps years earlier, in a learned set of rules, such rules are no longer a recognizable part of the expert's decision-making considerations.

Dreyfus' model is not meant to imply that experts act rashly or without deliberation. An expert may pause to deliberate before acting, but this is reflection on intuitive selection, not analysis. The rationality used by experts when they have time to ponder decisions is not calculative rationality (in other words reduction of the problem to elements for checking), but deliberative rationality. Deliberative rationality is 'thinking about the processes and product of his intuitive understanding'. (Dreyfus and Dreyfus 1986: 167)

If time is available, experts will consider the uncertainties around both their interpretation of events and the course of action they have chosen. Firstly, is there a completely different interpretation of the situation that would lead to a completely different appropriate action? The salient parts of the situation have been drawn into the foreground by the expert's past experience, but there may be grossly different interpretations of the situation that are possible. An expert with time to deliberate will look for such interpretations in an attempt to avoid tunnel vision. If time is available, an expert will also search for how the current situation differs from those experienced in the past and what that might mean for the course of action chosen. Is it possible to make allowances for unexpected outcomes that might come about as a result of the slightly different overall situation?

Dreyfus and Dreyfus' work on skill acquisition, originally published in 1986, has much in common with Klein's model of naturalistic decision-making (Klein 1998, 2003) described in the next section. It is apparently very relevant to time pressured decision-making by experts such as operational managers as we will see in later chapters, but it also has relevance for readers who wish to learn from this material. By seeing how the experts in these three organizations learn, what can we ourselves learn about safety decision-making? As described in the Introduction, Flyvbjerg (2001) proposes an answer to this issue for case study research in general by linking this type of research to learning by experts. Case studies such as those in this book provide, not new rules, but new stories that we can use to build our own expertise.

2.6 Experts and Sensemaking

As early as the 1950s,[3] it was recognized by some researchers coming from the social sciences sphere that real world decisions often did not fit the normative model favoured by the economists and behavioural psychologists described in Section 2.3. Apart from the issue of the cognitive effort required to evaluate possible strategies in the face of a specific decision, it was acknowledged that differences of opinion or simple uncertainty often exist about the actions required to achieve a specific goal, or even about the appropriateness of the goal itself.

A more recent view of decision-making research that focuses attention away from the rational actor model is the field of naturalistic decision making (NDM) (Klein 1998, Lipshitz et al. 2001a, Lipshitz et al. 2001b, Salas and Klein 2001, Klein 2003, Klein 2009). The underlying assumption of this work is that decision-making can best be understood and improved by studying expert decision-makers in naturalistic settings. Naturalistic decision-making research has shown that experienced people under pressure in complex situations do not generally use the classical approach to decision-making as described in Section 2.3. Under these circumstances, people tend to adopt what is termed the recognition-primed decision (RPD) model (Klein 1998). This work is based on field observations and interviews with fire fighters, neo-natal intensive care nurses, surgeons, weather forecasters, military field commanders and pilots.

In this model, decision-making is not a once-through process searching for the best option, but rather a cyclic process where the aim is to choose an acceptable option and then improve upon it, based on the observed system performance. The process starts when, in a specific situation, a decision-maker notices particular pieces of information or cues. The pattern formed by the cues is then recognized by the decision-maker based on experience. Also based on experience, the decision-maker focuses on a potential solution or action script and then imagines what might happen if this action were to be implemented. This involves experience again in the form of the decision-maker's mental model of the overall operations. If the imagined outcome is good enough then the action is implemented.

The process becomes cyclic as the situation changes, either as a result of the action taken or due to external influences. If the situation change is due to the action taken, then the change may confirm or challenge aspects of the expert's mental model. A decision-maker's mental model consists of the ingrained assumptions about the system and the cause and effect relationships between the various parts of the system. Mental models pervade the RPD process. This includes recognizing patterns of cues, their links to possible actions ('action scripts') as well as the likely

---

3   For list of references see – note 11 in La Porte, T.R. and Consolini, P.M. 1991. Working in Practice but Not in Theory: Theoretical Challenges of 'High-Reliability Organizations'. *Journal of Public Administration Research and Theory*, 1(1), 19–48. and note 1 in Etzioni, A. 1967. Mixed-Scanning: a 'Third' Approach to Decision Making. *Public Administration Review*, 27(5), 385–392.

effects of the actions. In a practical sense, the normative value of the RPD model is in emphasizing the importance of mental models. Improving the breadth and validity of mental models then becomes the practical strategy for improvements to decision-making.

Whilst Klein emphasizes the difference between this model and classical decision theory (primarily in that the recognition primed model does not involve comparison of options), there are some distinct similarities. Each model starts with problem recognition and definition and moves through selection and implementation of a course of action chosen either consciously by analysis or subconsciously, based on expertise. Some decision researchers, for example Laroche (1995), claim that the idea of decision-making in itself is a social construction and that this step occurs *after* the action has taken place to justify and make sense of our organizational experience.

All the fieldwork on which the RPD model is based has been carried out in an environment where feedback on the effectiveness of the decision is available in a fairly short (but not too short) time frame. The model assumes that adjustments to the course of action chosen are possible based on feedback received. This means that further information must be available to the decision-maker in time to adjust the chosen 'good enough' course of action and hence to improve the overall outcome. Conversely, if operations are tightly coupled (that is where events can escalate rapidly from the initial cues to an irreversible outcome) there may be no opportunity to improve upon an initial decision that was judged to be 'good enough'. Another factor in considering application of the model to safety decision-making in HROs is that, by definition, accidents are rare events and decision-makers may not be able to conduct an accurate mental simulation if their mental model does not accurately cover rare events in complex systems. It would seem that the RPD model has limitations as a strategy for complex, tightly-coupled systems.

A different view of the process of decision-making is offered by the literature on sensemaking (Weick 1995, 2001). Sensemaking 'is about the interplay of action and interpretation rather than the influence of evaluation on choice'. (Weick et al. 2005: 409) It is linked to NDM in that the process by which an expert updates his or her mental model could be described as sensemaking. This process is triggered when an unexpected or incongruous event occurs. Sensemaking describes the way in which organizational actors literally make sense of events and hence move to an appropriate action. It has much in common with the concept of mental models, but focuses on the process by which such models are continually formed and refined. This focus on process, rather than outcome, emphasizes the transient nature of sensemaking and the fluid nature of our interpretation of events. A decision-maker therefore no longer makes a choice, but acts deterministically as a result of the sense that has been made of the situation, at that instant in time when action is initiated. 'Order, interruption, recovery. That is sensemaking in a nutshell.' (Weick 2006: 1731)

Snook (2000) makes this point when he uses sensemaking as one frame for his analysis of the accidental shooting of US Blackhawks over northern Iraq (which resulted in 26 friendly fire fatalities):

> I could have asked, 'Why did they decide to shoot?' However such a framing puts us squarely on a path that leads straight back to the individual decision maker, away from potentially powerful contextual features and right back into the jaws of the fundamental attribution error. 'Why did they decide to shoot?' quickly becomes 'Why did they make the wrong decision?' Hence, the attribution falls squarely onto the shoulders of the decision maker and away from potent situational factors that influence action. Framing the individual-level puzzle as a question of meaning rather than deciding shifts the emphasis away from individual decision makers toward a point somewhere "out there" where context and individual action overlap. Individual responsibility is not ignored. However, by viewing the fateful actions of TIGERS 01 and 02 as the behaviours of actors struggling to make sense, rather than rational attempts to decide, we level the analytical playing field toward a more complete and balanced accounting of all relevant factors, not just individual judgement. (Snook 2000: 206–207)

In this view of organizations, decision-making becomes a retrospective process. Situational interpretation leads to action, which is then rationalized and described with hindsight as a process of decision-making. Laroche (1995) takes this view further in suggesting that decision-making is a social representation developed by managers who wish to take a heroic view of their own behaviour. He suggests that decision research has been limited by its failure to recognize the basic assumption that decisions and decision-making processes are realities. His view is that an action perspective is a more valid view of the reality of organizations and that decisions and decision-making should be studied as social representations that influence behaviour and understanding.

Sensemaking provides a useful conceptual link between the individual and organizational processes. Klein's work described earlier focuses on the experience of the individual as the key, determining factor in action selection. Other models portray organizational actors as victims of their organizational circumstances. The concept of sensemaking allows both aspects to be integrated into individual decision-making. This perspective also provides a framework that accommodates both rational and non-rational elements in that the decision-maker's actions are not seen as based only on logic and analysis, but also in the context of experienced judgement and organizational experience.

Weick (1995) details the following seven properties that distinguish sensemaking from other processes.

- Grounded in identity construction,
- Retrospective,
- Enactive of sensible environments,

- Social,
- Ongoing,
- Focused on and by extracting cues, and
- Plausible (rather than accurate).

By combining individual and organizational, rational and non-rational elements into a single descriptive framework, the sensemaking perspective provides a useful guide in consideration of the decision-making practices of operational managers.

2.7 Conclusions

This chapter has briefly reviewed six areas of research that are helpful in understanding how operational managers make decisions and, more importantly, how these practices may ultimately impact safety outcomes. High reliability theory highlights the need to develop organizational learning strategies that promote the ability to deal with unexpected events. These strategies include making the most for learning from any situation where outcomes are not as expected and valuing both deep system knowledge and diversity of views. Resilience engineering addresses operational decision making directly by inviting us to focus on 'sacrifice decisions' and by highlighting the need for both anticipation of problems and adaptation of response to abnormal situations that have already developed.

Other theories that address the subject of decision-making directly range from the rational, analytical approach favoured by classical decision-making processes to the flexible story-based approach found by the NDM researchers studying experts in the field. Rational choice theorists favour decision-making based on analysis and in the safety arena, this means risk management with its focus on consideration of frequency, consequence, controls, costs and benefits of a range of alternatives. NDM researchers instead focus on how experts draw on their experience when making time-pressured decisions. Theories of accident causation are less directly relevant in this normal operations study. Nevertheless, the Swiss cheese model highlights the importance of considering the organizational context to decision-making that is missing from either classical decision-making or NDM research.

Research on expertise and learning has emphasized the role of both judgement and application of rules, and how reliance on these aspects changes as experience increases. Perhaps the sensemaking perspective provides the broadest potential insight into decision-making by operational managers. This perspective connects the actions of the operational managers to their experiences and to their organizational environment.

All of these research areas provide relevant insights and understanding. The discussion of decision-making in later chapters is informed by all these views on safety and decision-making.

PART A
Decision-Making in Three High Reliability Organizations

This part of the book includes examples of decision-making by operational managers in three different organizations: a nuclear power station, a chemical plant and an air navigation service provider. For readers used to an exclusively technical emphasis, it may seem odd that practices in organizations that rely on such clearly differing technologies can be usefully compared. In fact these comparisons are valid because each can reasonably be classified as a high reliability organization or HRO.

As described in Chapter 2, an HRO is defined by the combination of an operating environment with conflicting pressures, a primary task that poses a significant hazard and an excellent operating record, including safety, production and socio-political dimensions. These three criteria are all met by the three organizations whose operational managers took part in this study. Before moving on to describing decision-making in each organization, it is perhaps useful to consider the overall organizational environment in which these organizations operate.

The first organization is a UK nuclear power station. The nature of the technology and the political circumstances under which the nuclear power industry has been operating for several decades in Europe and the US have resulted in a very strong focus on management of safety in that industry, and this facility operates under a safety case style regulatory regime which includes a requirement for the site to hold an operating licence. With a small number of well-known exceptions (Three Mile Island, Chernobyl, Fukushima), the industry record is good and nuclear safety practices and regulatory approaches have tended to be used as a model for other high hazard industries using complex technology. Nevertheless, the industry faces continuing government scrutiny as a result of public pressure. This particular organization also faces significant financial pressures. The facilities are nearing the end of their design life and within a few years will be shut down. The remaining productive life of the facility is being managed on a commercial basis by a statutory authority that has obligations under its foundation Act to return profit to the government. Cost and safety are both strong drivers for this organization and the impact of this on operational decision-making is described in Chapter 3.

The chemical plant is a well-established industrial site in Australia (operating since the 1970s, although ownership has changed several times since then). The site deals with significant quantities of several toxic and flammable chemicals.

The site holds a licence to operate under local regulations which were introduced following a major industrial incident in the State of Victoria that caused significant public outcry and resulted in a tough political response in the form of a new licensing regime. Good safety performance is therefore a socio-political necessity for this organization, driven by issues of corporate reputation, industry norms and regulatory requirements. The regulations are goal setting in style rather than detailing prescriptive requirements, and are generally similar in form to the regulations covering the UK nuclear power station. The plant also operates within the general pressures currently being experienced by the petrochemicals industry in Australia. Market pressure from China impacts both feedstock price (as competition drives feedstock prices higher) and product pricing (as new product capacity in China results in falling product prices globally). As will be seen in Chapter 4, operational managers working in this organization feel these conflicting pressures keenly.

On the other hand, the nature of air traffic control is somewhat different – provision of a service to a broad sector involving customers and stakeholders ranging from commercial airlines to sports aviators. Chapter 5 describes decision-making by operational managers in Airservices Australia. The role of the organization is to provide safe and environmentally sound air navigation services to approximately 11 per cent of the world's airspace on behalf of the Australian government. The organization is generically known as an air navigation service provider (ANSP). Separation of the provision of air navigation services from regulatory functions has been the policy of successive federal governments for more than a decade, and in 2007 the last remaining regulatory function (airspace management) was transferred from Airservices to the Civil Aviation Safety Authority. Globally, this is a common approach to management of air navigation services, with 38 ANSPs being established in the past ten years by transferring operational and financial responsibility for these services from national governments to independent commercial entities. This is the form that Airservices now takes (Airservices Australia 2007).

Since Airservices currently has a monopoly, the Australian Competition and Consumer Commission regulates pricing of the organization's core services. Consistent with government policy of increasing access to aviation, fees to customers have fallen approximately 30 per cent in real terms over the past decade. The financial model for the independent entity also has a goal of 25.3 per cent return on investment to the owner (the Australian government). Airservices' first priority has always been safety but the organizational focus has now been expanded in order to 'maintain world leading safety performance and improve the safety, efficiency and environment of the Australian air transport system'. (Airservices Australia 2007: 12) Government policy has set the organization the challenge of maintaining safety performance, whilst drastically improving its operational efficiency in order to cut costs to industry and return an operational financial dividend to the government. Operational managers at Airservices work in an environment of conflicting and changing priorities.

Based on the nature of their business activities, the commercial and regulatory pressures that they face, and their very good safety records, each of the three organizations could be characterized as a High Reliability Organization. It follows then that they might be expected to use similar strategies for managing safety in such challenging circumstances and therefore provide useful examples from which we can all learn. Despite their generally high level of performance, it is worth noting also that these three organizations are not perfect. Whilst many effective strategies for decision-making are identified, there are also a few situations described in the following chapters that could be seen as critical of these organizations in small but significant ways. This is perhaps particularly difficult in the case of Airservices, because they have a monopoly as the only air navigation service provider in this geographic area and so are clearly identifiable. These situations are described, not to find fault with any organization or individual for its own sake, but because of the valuable contrast they provide when compared with the vast majority of attitudes, behaviours and practices that support excellence in safety performance. Taking the approach of the HRO theorists, these small failures provide us all with important learning opportunities and these minor criticisms should be seen in that light.

The way in which operational managers in each organization make important time-pressured decisions is described in the following three chapters. Each chapter includes some background on the site and operating arrangements to provide context, followed by the heart of each case study – the stories told by operational managers about decision-making. This is followed by a summary of formal procedures in place, if any, for reporting and recording decisions and incidents and finally a description of overall attitudes towards decision-making in each organization. Each chapter concludes with a short summary.

Chapter 3

At the Nuclear Power Station – 'We Put a Line in the Sand'

The nuclear power industry has a reputation for having some of the most sophisticated systems in place for managing operational safety. Based on the practices seen at the nuclear power station described in this chapter that reputation is not undeserved although, not all activities were as controlled by procedure or as tightly specified as some managers imagined.

3.1 Site and Operational Background

The nuclear power station includes two nuclear reactors and has a total generation capacity of almost 1,000 megawatts. The physical plant and equipment, and the operations staff, are divided into two distinct technical areas – reactor equipment and power generation. In the reactor area, the energy from two nuclear reactors is used to generate large volumes of high-pressure steam. In the generation area, this steam drives turbines to produce electricity. Operational decisions in these two linked major groups of activities are of interest here.

This is a significant industrial facility. Approximately 650 people are employed at the site including approximately 170 people in the Operations Department. The operational arrangement includes three eight-hour shifts each day, with over 20 people per shift. As is typical for this type of staffing, there are five shift teams in total to fill the roster. Each shift includes 20 operators and two team leaders. The person with final responsibility for operating decisions on each shift is the Shift Manager. Research interviews therefore focused on Shift Managers and those with whom they consulted in making operational decisions.

The minute-by-minute operation of the plant is the responsibility of the Production Technicians. Many of the plant and reactor systems are fully automated whilst others require manual intervention to ensure they are functioning appropriately. The work of the Technicians is therefore a combination of performance monitoring, intervention, and routine operation of plant and equipment. Most members of the shift team spend their time in the plant areas. The plant and all work in progress are monitored, controlled and directed from a single control room. The control room is deep inside the reactor building and is permanently staffed by four people (the Production Team Leader and a Senior Production Technician for each reactor and for generation).

The Shift Managers report to the Operations Manager, who is a member of the site management team. The Operations Manager reports directly to the Site Manager, who is the most senior company representative on the site.

The Shift Manager has very limited involvement in the detailed operation of the plant when things are running smoothly. The Shift Managers share an office overlooking the control room. During afternoon shift or night shift the Shift Managers are mainly in the office or the control room itself. On day shift, they spend a significant amount of time in meetings away from the reactor building and plant areas. The Shift Manager on duty is always immediately contactable via the plant radio system or a pager. His attention to the operational details increases when something out of the ordinary is planned (such as a startup or shutdown) or when a significant operating anomaly develops.

The nuclear power station has experienced one significant nuclear safety incident over the operating life of the facility. In the early 1990s there was a significant incident on site. Although it did not result in any radiation release, it was judged to be an incident that could have resulted in a significant radiation release and hence was reportable to the authorities. The company was fined by the regulator and the site received a lot of negative publicity. It is well remembered by the staff, as most have been there since before that time. It was mentioned by some staff in interviews and was cited by the Site Manager in a site-wide meeting as something that no-one wants to see happen again.

More recently, the site has had an excellent record in both nuclear safety and traditional industrial safety.

3.2 Operational Decision-making

Complex plant such as the nuclear power station operates in a quasi-steady state, where minor equipment faults and small deviations in the process conditions occur constantly. The status of major equipment items (such as pumps, compressors and major valves) is shown by lights on the control room panel and a change in status causes an alarm (both audible and visual). If, for example, a pump stops automatically due to a problem detected by its control instrumentation, then an alarm sounds in the control room and the status lights move from on to off. Similarly, many process parameters (such as pressure, temperature and flow) are measured and displayed on control room panels (in the form of paper strip charts or computer screens). Deviations from the expected value cause alarms when pre-defined levels are reached.

In general, it is the job of the Production Technicians to analyse the causes of process deviations and minor equipment problems and to initiate changes to plant operation as required. Shift Managers become involved only with process deviations that are significantly out of the ordinary.

In this operating environment, the operational decision-making in the face of a developing operating anomaly followed a clear two-stage process.

3.2.1 Stage One – Compliance with Station Operating Instructions

Firstly, when an abnormal operational situation with potential safety implications is initially discovered, the Shift Manager makes a decision about the need for immediate action to bring the system to a safe state. This usually (but not always) means deciding that a reactor shutdown is necessary, despite the attendant interruption to power generation. This step was often dictated by application of formal rules in the form of Station Operating Instructions.

A decision to take a reactor off line then becomes one of compliance with the relevant Station Operating Instruction. Shift Managers were very well aware that a compliance breach (in this case not shutting down a reactor when such a shutdown was required by the formal rules) was considered to be a major disciplinary breach with potentially serious career consequences. As described in Section 3.4, Station Operating Instructions derive their authority directly from the statutory operating licence of the facility. Story 2 describes a case such as this.

STORY 2: WE SHOULD HAVE A MARGIN

Shift Manager Interviewee 1 told the story of a delayed start up due to concerns over getting close to the limit imposed by a Station Operating Instruction. On night shift, the operating crew was preparing to raise the reactor control rods (in preparation for starting the nuclear reaction) when the on-duty chemist came into the control room. Her monitoring activities were showing that one process gas parameter was approaching its occupational health limit, that is the limit specified in the Station Operating Instructions. The Shift Manager continued the story as follows: 'So, I decided then that I was not happy to continue that close to ... although we were within, I wanted a margin. So I consulted offsite with the various standby personnel who agreed that we should have a margin, so I shifted the focus of the team away from starting up and said, "No we're not starting up for at least a day. Let's get some margin and get back tomorrow." [This meant] about a 24 hour delay. But you know, again, if there had been a lot of pressure for production we could have in fact gone, because we were within limits, but it was too close to it for my liking and there was full support.'

The Production Technicians also make decisions about whether a particular situation complies with this set of rules and may act without consulting the Shift Manager, depending on the available time for decision-making and the nature of the potential breach of the overall operating envelope. Story 3 is an example of a case where the decision to shut down the reactors and turbines was made without checking with the Shift Manager in advance. Despite the major consequences (in terms of both production and workload for the team) the decision was driven by a clear compliance requirement.

STORY 3: YOU NEED TO BUTTON YOUR REACTORS AND BUTTON YOUR TURBINES

Shift Manager Interviewee 2 was out in the plant when the sudden loud noise and visual impact of high flow in the venting and relief system made him aware that a plant shutdown had started. Returning to the control room, he discovered that a full reactor and turbine shutdown had been initiated by the control room staff. A failure had occurred in the low voltage power system, which meant that the Production Technicians had lost access to the data presentation system that allowed them to monitor the plant. This system is separate from the plant automatic control system so the plant could continue operating, although the Production Technicians would not be able to 'see' what was happening. Station Operating Instructions call for an immediate shutdown initiation in these circumstances. The Production Technicians apparently saw this as a clear compliance issue and did not feel that they needed to consult the Shift Manager before taking the step of commencing the shutdown. The Shift Manager was in complete agreement with the actions taken by his team.

Shutting down the reactors and turbines initiates a period of very high workload for the entire shift to shut down and isolate all equipment and perform all necessary external notifications (since power generation has ceased).

In this case, control functions were not impacted by the failure, so, in theory, operations could have continued without monitoring until the data presentation system was restarted (estimated to take 20–25 minutes). The on-duty staff were well aware that Station Operating Instructions require an immediate shutdown if monitoring is lost, even if there are no signs that the reactors and plant are not operating normally. This was the action that they took.

If there is more time available to make the decision (in other words there is no immediate clear breach of a Station Operating Instruction) and there are broader implications for the action (for example, shutting down a reactor, rather than just one section of plant), then the final decision will be made by the Shift Manager. This is partly due to the production interruption involved, but also because there are hazards associated with an unplanned shutdown of the system. For both reasons, the decision to shut down a reactor is certainly not taken lightly. Nuclear reactors produce enormous amounts of energy which must be removed from the reactor in a safe manner, even if the energy is no longer to be used for power generation. The physical shutdown of the nuclear reactor itself (planned or unplanned) is the job of the licensed control room technician. The required sequence of activities is complex and much of the sequence is manually (not automatically) initiated based on the technician's judgement that it is safe to proceed to the next step. These sequences are practised in a plant simulator, and the demonstrated ability to manage such events is a key part of a technician's licence testing. The Shift Manager has no direct operational role once the decision has been made that the shutdown is required, unless he needs to manage broader aspects of the incident that resulted in the decision to shut down in the first place.

There are other reactive cases where the Shift Manager decides that a situation is so abnormal that an immediate shutdown is required, even if there is not a strict breach of a formal rule. As Interviewee 2 described it: 'if there are enough pieces of the puzzle missing and I can't say hand on my heart that I know we are safe at the moment then I need to go to some arrangement, status, condition where I do know that we are safe. It may well mean shutting the plant down.' Within the operating envelope represented by the formal rules, Shift Managers rely on their own experience, and that of their colleagues, to make these types of operational decisions. The result may be a decision to shut down part or all of the system, even though the plant is within its formally defined operating envelope. Story 1 (in Chapter 1) describes a case like this. Story 4 is a similar case that involved decreasing power generation, rather than shutting down a reactor.

**STORY 4: IF ANOTHER FAULT DEVELOPS, THE MACHINE MIGHT
NOT SHUT DOWN**

When acting as Shift Manager, Interviewee 5 was reviewing the control room alarms that were up at the start of his shift one Sunday evening. One alarm indicated that a primary coolant circulator had stopped due to a fault but, in fact, the equipment was still running. This seemed to suggest that there was a problem of some kind in the control system for this piece of equipment. The control room team investigated the cause of the alarm and checked the engineering drawings to make sure that the fault that had occurred should, indeed, have caused the circulator to shut down. This was the case.

After consulting several specialists, Interviewee 5 decided to shut down the circulator which required a decrease in reactor power and hence a drop in electricity generation. He described his reasoning as 'we made a conservative decision to take the circulator off because we didn't have any confidence that the circulator would have tripped in the event of low lube oil or seal oil pressure. So that's what we did, we took it off'.

The primary coolant circulators play a key role in removing energy from the reactor core as is necessary for power generation purposes, but also to keep the nuclear reaction under control. The risk of running the equipment with the control system fault is that, if some other fault developed (such as low lube or seal oil), then the equipment might not be shut down automatically in a controlled manner, but run until it failed in some way. This would cause damage not only to the circulator itself, but would mean a much less controlled change over to secondary and tertiary systems for reactor cooling.

The next day, maintenance personnel investigated and repaired the minor fault which had caused the malfunction and the circulator was restarted 24 hours after it had been shut down. Power generation returned to full capacity.

3.2.2 Stage Two – The Line in the Sand

The second stage in the decision-making process comes about if the supervisor decides that no immediate shutdown is required. In this case, operations can continue, but with the situation being closely monitored due to a reduced safety margin. Whilst the safety margin is reduced from the normal and desired level, it has not yet reached a situation that the Shift Manager sees as unsafe. This might be an equipment breakdown with some associated down time required for repairs or an unusual operating condition that needs further investigation. Personnel at the power station developed and imposed limits on operation under these types of conditions. The relevant limit is often, but not always, time, as was the case in Story 5 below.

As another interviewee described in a story about the timing of repairs: 'We put a line in the sand. If it's not fixed in the second week, we're coming off'.

STORY 5: NOT A DECISION YOU MAKE LIGHTLY

A water leak from service piping developed high up in the reactor building. Whilst the leak was not hazardous in itself, the large volume of escaping water had the potential to inundate electrical equipment lower down in the building and perhaps cause serious problems.

Initial efforts by the operating team focused on controlling and repairing the leak and the standby maintenance repair team had been called for assistance. Shift Manager Interviewee 2 described the situation: 'Have we met any specific criteria for taking this reactor off in our technical specifications? No we hadn't ... So it was a case of making a decision. OK enough is enough ... I am going to set myself a line in the sand beyond which I need to do something ... It got pretty close. My criteria was I could allow this to go on another 10 or 15 more minutes ... or any other plant signal of distress ... if we hadn't sorted it out we would have been taking the reactor off. Again that's quite a heavy burden too. On the other side of it you know the implications, work-wise, safety-wise, cost-wise as well, if necessary as part of that. It's not a decision you want to make lightly, but it's something that goes with the territory if you like. That's one of my calls. To be able to make that call and say yes I was not comfortable in that situation hence I need to take this conservative decision to shut the reactor down so we can do a repair to it.'

Another case where time is the key parameter is given in Story 6. In this case, the time constraint applied to technical advisers rather than the maintenance crew.

STORY 6: SHOW ME WHY IT'S SAFE TO CONTINUE

Each fuel rod in the reactors is contained within a standpipe. Each standpipe includes a special type of window that is used during refuelling. Shift Manager Interviewee 1 recounted one occasion when control room staff conducting routine refuelling activities noticed (in the images received from their monitoring camera) what appeared to be a crack in one of the pieces of window glass. The glass is remote from the operators, within the radiation shield of the reactor itself.

The concern of the control room crew and the Shift Manager was the potential implication of failure of the window, although none of the 400+ equivalent windows had failed in the 30-year operating history of the facility. They were concerned that pieces of glass might interfere with the flow of coolant around the fuel rods – a significant safety issue. On this occasion, the Shift Manager contacted specialist staff (the event occurred outside of office hours) and explained the situation. 'So what I said there was – You show me, very quickly, that we don't need to shut down.'

The specialists went to work and within an hour produced a written technical argument about why it was safe to continue running the reactor. The Shift Manager was convinced by the argument and operations continued.

Sometimes other additional parameters are also called into play, as in Story 7 below.

STORY 7: AN UNUSUALLY STILL DAY

One of the Shift Managers (Interviewee 7) recounted the story of preparing the plant for a planned outage. This involves venting a large volume of carbon dioxide from a vent stack designed for this purpose. The coastal location of the plant means that the normal weather pattern is very windy, but on the day the plant was to be shut down and vented the air was completely calm.

The Shift Manager had never come across this situation before and there was no mention of limiting weather conditions in the Station Operating Instructions. Nevertheless, he was concerned that, in such still conditions, the carbon dioxide might not disperse adequately and might fall to the ground in and around the reactor building as carbon dioxide is heavier than air. In the worst case, this would be a safety issue as a potential asphyxiation hazard. He considered delaying the plant preparation, but the weather forecast was for continuing still weather. Instead, he chose to temporarily de-staff parts of the reactor building that he thought might be vulnerable and to put in place a system of temporary carbon dioxide monitoring so that venting could be stopped immediately if elevated carbon dioxide levels were measured at ground level.

In fact, the activity proceeded without a hitch, but the Shift Manager felt completely justified in having delayed the work until he could put an extra barrier in place just in case.

These limits are articulated explicitly to the Production Technicians and other control room staff. Many cases were recounted by interviewees where the limit was reached and shutdown initiated (as well as many cases where the problem was fixed within the self-allocated window, for example Story 8).

STORY 8: WE SET BOUNDARIES AND WORK WITHIN THOSE

One of the Shift Managers (Interviewee 1) recounted the story of restarting a reactor after a planned outage. Part of the way through the start up sequence, the control room Production Technician found a problem with moving the control rods (one of the key devices used to control the reactor power level). The Production Technician's initial response (shouted out to Interviewee 1, who was in his office adjacent to the control room at the time) was to plan to shut the reactor down again immediately.

Further investigation (over a few minutes) showed that the technical problem was such that it was possible to manually decrease the power of the reactor but not increase it. Interviewee 1 described this as 'partial control'. It was also established that, whilst there is a minimum power level specified in the Station Operating Instructions, they were well above that figure. Before continuing, they set themselves a limit. 'We gave ourselves a bound of power level whereby, if it went down so far towards the automatic trip, then we would have tripped it anyway ... So, we set boundaries and worked within those.'

Within two to three minutes they had solved the problem and were back to power raising.

All three organizations studied operate some form of maintenance works approval system where specific authorization is required from the operations supervisor for equipment to be taken off line for maintenance. If the equipment involved has a safety function, these types of decisions also require a trade-off between safety and production or cost. Unlike the operating situations described above, these decisions are made proactively in that the supervisor can choose the time at which the work is done.

At the nuclear power station, the Station Operating Instructions include a listing of minimum equipment that must be available at all times for systems with a safety function. This is used as a conservative guide for determining whether equipment can be taken out of service for maintenance. Story 9 is an example of this.

STORY 9: CONSERVATIVE MAINTENANCE PLANNING

Shift Manager Interviewee 2 described how the operations group decides whether equipment should be released for routine maintenance. 'If we've got five items of plant and we only need two of them [to meet the requirements of the Station Operating Instructions], we are not going to release three of them. You may release one or we may occasionally give them a second one to work on, but we're certainly not going to give them three at the same time, even though the specification says [we only need two], because it's too close to the wire.' Note the very conservative approach here. The operations staff will not deliberately operate the reactor at the safety limit, but will always maintain a margin if they can so that an unexpected equipment failure still leaves sufficient redundancy in the operating system for them to continue running.

Interviewee 2 goes on to describe his thinking behind the decision-making and the tension this brings into his relationship with the Maintenance Department: 'We're trying to maintain the conservative side of it. We've got standby plant and it's only due to poor planning and not configuring outages at the right time. They are reducing the buffer, the safety case, the barrier to the worst event that may happen. I would much rather have a longer period with only one gas turbine unavailable rather than a shorter period with two or even three unavailable. That's that side of it. So sometimes we get into a battle saying we're not going to give you a piece of kit until you give us that one back.'

These nine stories (eight recounted in this chapter and one in the Introduction) typify the conservative tone of the interview data generally. Operating outside (or even within, but close to the boundary of) the defined operating envelope was seen by all Shift Managers as unsafe. Deliberately putting or allowing the plant to be in such a condition was almost unimaginable to the Shift Managers and their operating crews. On the other hand, operations within the defined boundary were not always seen as necessarily safe. The Shift Managers accepted that, as a normal part of their job, they would come across unusual operating combinations that had not been considered by the people who developed the operating envelope. In such cases it was up to them to decide whether production could continue safely and to seek whatever specialist advice they felt they needed to make that decision.

3.3 Reporting and Recording Decisions and Incidents

Many organizations, including this nuclear power station, have comprehensive systems in place to record incidents. Part of the objective of these systems is to allow organizational members to learn from past incidents to prevent recurrence. When judgement forms such an important part of safety decision-making, this seems like a robust strategy. In fact the site-wide incident reporting scheme in place at the nuclear power station has little overlap with the types of incidents and occurrences of interest to the Shift Managers in enriching their operational knowledge. The only exception to this was incidents that had the direct potential for loss of some kind. This is discussed further in Section 8.3.

More commonly, the types of incidents described above were recorded on forms stored in the Shift Managers' office. They are colloquially known as QS24 forms. Apart from identifying details for the originator, the form has only five fields:

- event title
- description of event
- persons contacted
- result of discussion
- decision

The reports (approximately 300 from an eight year period) are kept in paper copy only and are used by the Shift Managers as one means of sharing their experience. The existence of these records was not widely known outside the group consisting of the Shift Managers themselves and the technical specialists whose conversations are recorded on these forms. Certainly none of the Safety Department staff was aware of this method of recording safety decisions or the existence of the files.

Interviewee 1 spoke about the purpose of the forms as follows:

> It's to record advice or discussions so that if we end up in court or an inquiry down the line I can show that, I can demonstrate, I have a record that yes I did consult these personnel. Yes, we did consider these things and we did reach this decision. If it later proved to be flawed at least there's a record that it was made in good faith at the time. But it also helps in formalizing it slightly. It does mean that you have to think a bit more carefully about what you are doing.

He also said:

> There is [a lot of good experience in there]. It's handy you know. Occasionally there will be something and you think I'm sure that's happened before and you can look in here and there might be something.

Interviewee 2 spoke about QS24 forms:

> If we can't write down the rationale as to why we are doing something it probably means that the rationale is suspect to start with. It makes you go through that process, formalises that. It also means if someone challenges at a later date "why did you do that?" well this was the reason at the time, these were the conditions on the face of it, this was the discussion that took place and this was the outcome that we agreed to.

Interviewee 4 described three reasons for using the QS24 forms.

> One, obviously there's a lot going on in your mind at the time. You've got to try to concentrate on what the guy is telling you. You might misunderstand a certain sentence, or it can be down to a comma or a full stop and it might mean something different. So by listening and giving it a verbal ten minutes later, it is already second hand. I am telling somebody else – now it's third hand. There's a good potential for messing it up in that respect. So it's important to get it written down so it's always first hand.

> Secondly, the individual is hanging his hat. He then has a responsibility to give sound advice. Now if I misrepresent him in the fact that what comes out of my mouth is different to what he told me in my ear, then I need to make sure that that individual is happy with his advice. So by going through and reading what I have written to him and he hears what I've just said he can say hang on that's not right and equally when he is happy with it I can put his name and my name. It becomes very important then and has the potential obviously to be used in any inquiry or court even if it gets to that level.

The third reason was: 'if it is a long drawn out thing, the guy who comes in this evening, tonight, if there's anything that's going on well then there's your information. Have a read and then he asks me questions on the information'.

Technical Specialist Interviewee 6 said:

> The logic that we follow in these situations is that we start the consultation and we follow it through and we record our logic. If we are going to keep the reactor on, we've got to be able to write down why. Some people get this bit a bit wrong actually. They think the reason we write down why we keep the reactor on or whether we take it off is some kind of legal record so that if anyone comes afterwards they will be able to say this was why we did it. That is one of the reasons, but the real reason we do it is to test the logic really. It's very easy to sit around the table and discuss something and somebody say … oh yes that seems sensible or whatever. But the real test is if you can write down the logic of why you are doing something.

In summary, all Shift Managers could see the value of the main site incident reporting system and actively participated in various parts of this process. Nevertheless, when it comes to recording their own experiences they used a separate system designed to meet the needs of the small group of people who share that position. The QS24 form no doubt has bureaucratic origins within the station management system, but no one in the Safety Department was aware of these records. Their existence had been forgotten by all but the Shift Managers who have used this system as the basis of their own records of experiences and decisions made. Their view of the system of recording decisions was partly the cognitive benefits but also linked to compliance and legal justification.

As the following section shows, parts of the wider organization also had a strong focus on compliance with safety rules, but the attitude towards the role of experience and judgement was less uniform.

3.4 Organizational Attitudes Towards Safety Decision-making

The Shift Managers had a very strong cultural norm amongst themselves as to how operational decision-making was carried out. This way of working had been developed amongst themselves but organizational safety theory reminds us that they are likely to have been heavily influenced by the organizational context within which they work. Broader views on safety decision-making are therefore relevant. There were two very different approaches to thinking about work promoted on the site.

A view of the historical attitude to work is provided by the site induction – a half-day, competency-based program for all newcomers. The first item on the training agenda is site licence awareness and all attendees are given a copy of a booklet, which contains the full text of the licence (over 20 pages) and a 50-page summary of the site's Management Control Procedures (the quality system that implements the requirements of the licence). The prime message communicated in the training was that everyone who enters the site must obey absolutely all rules and procedures. This was repeatedly emphasized and trainees were told that, since the station rules and procedures form part of the site licence (issued by the government regulatory agency), breaking any rule or procedure was against the law and could leave an individual open to prosecution. The need to obey the law was even linked by the trainer (a retired long-term employee) to the presence of the armed police officers who were continuously present at the front security entrance to the facility!

This prescriptive and rather punitive attitude to people (both employees and visitors) in relation to their safety responsibilities was in sharp contrast to the views expressed by the current Site Manager, who had been in that position for only a year. In presentations to staff and material posed around the site he encouraged an orientation towards innovation and imagination along with an attitude of personal responsibility. He emphasized the need for all employees to 'make a personal choice for excellence' in all that they do. What he meant by this in practice was not articulated explicitly, but seemed to include a combination of conservatism, seeking to learn from mistakes, innovation and imagination.

The need for innovation was being promoted particularly strongly at this time. This message was communicated around the site by posters which included quotations from a broad range of sources to promote this style of thinking. One quote from Woody Allan says: 'if you are not failing every now and again, it's a sign you're not doing anything very innovative'. This is an attitude of risk-taking rather than one of compliance, which seems at odds with the very conservative attitudes of the Shift Managers. Whilst the message may be somewhat surprising in the context of nuclear operations, it seemed to be aimed at the change in management attitudes required as the site simultaneously reaches the end of its design life, moves into decommissioning and the facility changes from being a government-owned utility to a site for sale to the private sector with a strong commercial focus. The company that owns the power station had a corporation-wide campaign running at the time to generate new business ideas 'that will bear the hallmarks of simplicity, clarity, grace and beauty'. The range of transitions on the horizon requires people to 'think differently', both in business and technical terms.

In summary, the senior management view of safety decision-making had apparently changed from one based solely on compliance, to a more sophisticated view that acknowledged the place of judgement and experience. On the other hand, the organization was also under pressure to move to a more commercial focus and manage major upcoming externally-imposed changes. The management approach proposed to deal with this was to encourage innovation and imagination. Perhaps the site manager was expressing his personal conundrum when he made the decision to have the centrally controlled screen-saver for every computer on the site pose the question 'How can we innovate, inspire and imagine without losing our safety margins?'

3.5 Summary

Operational decisions regarding safety at the nuclear power station are made within a highly disciplined organizational environment. This is reflected in the two types of decisions made to shut down the reactors (or other major plant). In the first instance, operating limits or boundaries in the form of the Station Operating Instructions are very strictly adhered to. Secondly, even within the operating boundaries, some situations may be judged by Shift Managers, based on their experience, to be not acceptably safe. Shutdowns are initiated in some circumstances that are not specifically prescribed by the Station Operating Instructions if the Shift Manager is not completely certain that the situation is sufficiently under control.

In an abnormal operating situation where a shutdown is not seen as immediately necessary, a second discipline comes into play – conservative decision-making. In practice, this has been developed into a system whereby individual Shift Managers in a given situation may set a context-specific line in the sand. The line in the sand is a limit fixed in advance to define the point at which a shutdown will be instigated. This line in the sand is articulated to the entire operating team and examples were given where this self-imposed limit, once set, was treated as seriously as the Station Operating Instructions. This approach replaces an ongoing need for judgement calls under pressure with a absolute limit which has been determined based on the specific circumstances and tested as thoroughly as circumstances allow.

The organization operates a comprehensive reporting system to capture all kinds of incidents in an attempt to learn from them, but this is largely irrelevant to operational decision-making about safety. The Shift Managers are a very experienced group and yet, if there is time available, they prefer to consult their colleagues on decisions to be taken. Even such consultation has been proceduralized to the extent that the results of conversations are recorded and filed. It is these records that are most valued by operational staff as reference for learning about operational incidents that have occurred and decisions made to deal with them.

In organizational terms, the formal systems in place at the power station about safety relate primarily to compliance with rules. There is a highly developed, regulatory-based process for developing and formalizing operating instructions. Senior site management, certainly historically, has seen compliance as the prime method of ensuring that good operational decisions are made. Conversely, the focus on conservative decision-making is an implied acknowledgement that not all cases are covered by the published rules. If all decisions were formularized in procedures, such an exhortation would be unnecessary.

The senior management view of safety decision-making appears to have recently become more sophisticated, acknowledging the role of experience and judgement. The most recent message from senior management was that cognitive qualities such as innovative and imaginative thinking are important factors for organizational success in an increasingly competitive and commercialized environment. The best way to balance these two factors with the need for conservatism in safety decision-making was an open question at the time of the research.

Chapter 4

At the Chemical Plant – 'If it's Not Safe, We Don't Do it'

The global chemical industry has a mixed record when it comes to operational safety. This site is located in Australia where process safety has been the subject of significant government attention, particularly since 1998. In that year a fire at the Longford gas processing plant in Victoria led to two deaths, but also caused significant restrictions to domestic gas supply in that state for several weeks. The Royal Commission that resulted provided the political impetus to introduce safety case style regulation for onshore facilities (Dawson and Brooks 1999). This step change in the level of regulatory attention to plant safety issues has undoubtedly also increased industry attention to this area.

The site described in this chapter has other significant internal and external pressures. There have been several changes of ownership in the preceding decades. In addition it operates in direct competition with facilities in China which compete for both feedstock and customers.

4.1 Site and Operational Background

The chemical plant uses feedstock and creates products that are highly flammable and also toxic. The product is manufactured in several reactor vessels called autoclaves. The chemical reaction that converts the feedstock to product also produces energy and would accelerate if not controlled by removal of heat. The reaction section of the plant operates by converting batches of feed into product. Other parts of the plant operate on a continuous basis and are devoted to feedstock preparation and product recovery.

Over 100 people are employed at the site. Similar to the nuclear power station, the site is staffed on a 24-hour, seven days a week basis by five shifts. The Shift Manager has ultimate responsibility for operational decisions made on shift and is the most senior person on site outside normal business hours. Interviews focused on Shift Managers and those they consulted in making operational decisions.

Each shift also includes five Production Technicians – one in the control room, one in each of three designated plant areas, plus a fifth person to cover leave and training requirements. The control room is located close to the plant area. The Shift Managers have a small dedicated office area within the same building and spend much of their time in the office, in the control room itself or outside in the plant area. During normal business hours they spend significant time in meetings, either in the workshop area close to the plant or the administration building located further away outside the plant fence. They are continuously available to plant personnel via the plant radio system. The level of hands-on involvement that a Shift Manager has in plant operations at any given time depends on the level of experience of the Control Room Production Technician on shift and on the level of activity.

There are six Shift Managers in total to cover absences such as holidays plus one day shift position filled by a Shift Manager on assignment that covers general operations co-ordination, so there are seven people in total at that level. The Shift Managers report to the Production Manager, along with all other department heads for functions directly associated with product manufacturing (such as engineering, maintenance and laboratory services).

4.2 Operational Decision-making

Whilst smaller and less complex than the nuclear power station, operations at the chemical plant are similar in many ways. The flow of various chemicals is monitored electronically and the current status of a wide range of operating parameters (pressure, temperature, flow) is shown on plant instrumentation. The details are always changing, partly because of the quasi-steady state nature of such complex facilities and partly because, in this case, the heart of the plant operates on a batch basis that is the product is produced in batches (rather like baking a series of cakes), rather than in a continuous process.

At the chemical plant, some developing operational situations resulted in an immediate decision by the Shift Manager to shut part or all of the plant as described in Story 10.

STORY 10: PLANT SHUTDOWN FIRST AS LAST

Some parts of the process operate on a batch (rather than continuous) basis and there is intermediate storage capacity between some units. This means that when an equipment item breaks and part of the plant stops unexpectedly, decisions must be made about which other units should be shut down and at what point.

Shift Manager Interviewee 1 told stories about two recent occasions on which equipment failures occurred on his shift. On both occasions, he chose to shut down the entire plant once the nature of the problem became clear (only a matter of minutes after the problem came to light in each case), rather than continue production. The alternative course of action would have been to run parts of the plant on a piece-meal basis to fill intermediate storages in an attempt to try to minimise the impact on production. This has potential safety implications due to the unusual operating modes involved, especially occurring at the same time as operating staff are trying to resolve the initial problem. The intermediate storage capacity can also be an important buffer if there are problems on start up.

The safest and most conservative course of action to ensure production is not interrupted in a serious way is to shut down the entire plant sooner, rather than later.

Shift Managers all saw such decisions as being based on their judgement and experience, and not on application of formal rules and procedures although in some case it was arguable that the rules were simply the formulation of the collective experience as described in Story 11.

STORY 11: MINIMUM MANNING LEVELS

Shift Manager Interviewee 1 recounted the story of an occasion where, through illness and lack of availability of a replacement Production Technician, only three Production Technicians were available to work a shift in the plant instead of the usual four.

In practical terms, it is possible to run in the short term with only three Production Technicians but there is insufficient capacity to manage any maintenance work or attend to developing operational issues. Interviewee 1 said that he chose to shut down one section of the plant that is relatively simple and hence safest to restart. He described the decision in practical terms based on the tasks that needed to be done for the plant to run safely.

The Safety Case specifies that the minimum number of people necessary to safely run the plant is four. Running with fewer than four is therefore likely to be seen as a serious OH&S breach and could result in a formal Improvement Notice under the OH&S Act if discovered by the regulator. This limit was set when the Safety Case was prepared based on previous operating experience. Interviewee 1 continued to describe this situation in terms of experience of what was appropriate rather than compliance with the rule.

Some interviewees took exception to the way the research question was framed and claimed there was no conflict between safety and production, as safety is always number one. The following are typical statements from interview:

Safety health and environment issues, the plant comes off. Depending on the severity of the issue, the plant comes off. It's straightforward. (Shift Manager Interviewee 1)

If it's going to cause anybody harm or potentially cause anybody harm we won't run it. We won't do it. (Shift Manager Interviewee 3)

Safety or production, production loses out. (Shift Manager Interviewee 5)

Whilst there was no collectively articulated view about what specifically constituted safe or unsafe, the common view was that all seven Shift Managers had a similar understanding of what constitutes safe (in contrast to some points in the past when Shift Managers reported that views had varied significantly). This common conservative attitude had evolved over some years with strong support and emphasis from senior site management. One key point recounted by several interviewees was the story of the behaviour of a previous Shift Manager as described in Story 12. Some interviewees were uncomfortable with the apparent dismissal of the person involved, but everyone who mentioned the incident believed that the individual's behaviour was unacceptable and that it was a good thing overall that senior management had also made that clear in their response.

STORY 12: A JOKE GONE WRONG

Shift Manager Interviewee 5 told the story of a Shift Manager who had worked at the site some years previously and who he felt had had a poor attitude to safety. From the perspective of the operating staff, things came to a head with an incident where the Shift Manager visited a location on an elevated platform where maintenance work was being done. The person telling the story estimated that the work location was at a height equivalent to a six storey building. The Shift Manager picked up a loose blind flange and tossed it off the platform, apparently as a joke.

Someone in the general area below where the work was being done saw and heard the flange fall to the ground. Luckily no-one was hit and no damage was done to the plant, but an item like that falling from a significant height could have caused a fatality or serious injury, or a significant leak of toxic and flammable material from a damaged instrument fitting or small bore line.

The interviewee recounted that, after a formal investigation, the person involved was no longer employed at the site.

No interviewee was able to articulate any analytical process as to how he came to a conclusion about whether a particular situation or course of action was safe or otherwise. Most interviewees said it was based on their experience and judgement and several then told stories as to how they developed their sense of what is safe.

Shift Manager Interviewee 6 said, 'I guess like any decision you use your frame of reference in terms of what happened in the past and what I've seen in the past and what have I seen other people from other industries or other chemical plants or … mining plants [do]. You take all that into account.'

Although they were not directly asked about the incident, several interviewees recounted in various ways the story of the most serious incident to have occurred at this site approximately seven years earlier. For many interviewees, this incident taught them the potential of the plant for serious safety consequences. The incident resulted in a significant release of flammable (and toxic) material, but it did not ignite and no one was injured, although there was a full call out of external emergency services. The material was released from the process in an unexpected and complex way that took some time to diagnose and bring under control. Shift Manager Interviewee 7's summary is typical of the way the incident was described.

> I was involved here maybe five years ago now, where we had an incident which was potentially quite hazardous... I think at the time your training kicks in and you just think about making the plant safe. Afterwards you just think what could have happened…and you think what happened at Longford...[1]

Whilst Interviewee 7 was present during the incident (which occurred on night shift), the sense of shock at 'what might have been' extended to others who worked at the site at the time. One worker (Interviewee 9) described how concerned he was to arrive at the site for day shift to find a large number of fire trucks already there.

Interviewee 5 was involved in the incident as a Production Technician and he also recounted the technical sequences of events in detail, including the complex activities that the team went through to find the source of the problem. The punch line of the story for him was:

> Sometimes you actually do need to step back. You can't rush in. That's where [the former Shift Manager who features in Story 12] always came unstuck instead of working out, sitting down each time and doing it slowly. I guess you need to take a bit of time, step back and look at the overall picture rather than just being narrow minded. Some people will have a single view. It's better to have an overall view as a Shift Manager. I guess it's my responsibility to take an overall view.

At the chemical plant, the decision to stop or shut down the plant can have complex consequences that lead to safety issues and further safety/production trade-off decisions. Story 13 is one such example.

---

1   The fire and explosion at the Longford gas plant in Gippsland, Victoria in September 1998.

STORY 13: DAMNED IF YOU DO...

Shift Manager Interviewee 3 told the story of a significant production interruption. The incident started with a noisy stirrer in a reactor. The level and type of noise being made by the mechanism was such that the operating crew was concerned that a seal could fail, resulting in a major leak of flammable and toxic chemicals to atmosphere. The Shift Manager made the decision to stop the stirrer and abandon the batch of product that was being made.

The next issue was how to dispose safely of the partially processed batch. There are several safety systems in place to assist in this process, but the appropriate choice is specific to each individual case and needs to be decided by the operating team at the time. In this case, the Shift Manager made a series of technical decisions about how to respond, with the aim of safely dealing with the partially-reacted contents of the reactor whilst minimizing the production disruption and getting back on line as soon as possible.

In hindsight, some of these were misjudgements and the final result was a reactor jammed solid with polymer that had to be physically removed from the reactor vessel. To achieve this, the vessel had to be opened and the solid plastic contents jack hammered out. This posed a significant occupational health and safety challenge due to the potential for exposure to chemicals and difficulties in safely preparing the blocked equipment for work. The total interruption to production was 14 days, of which only four days related to the original problem with the stirrer. The remaining ten days were needed to manually clear the hazardous solid material from the plant.

For the cases in which no immediate shutdown was deemed to be required, supervisors responded in a similar way to those at the nuclear power station.

The equivalent parameters (such as estimated time to repair) were taken into account in deciding how to proceed as illustrated by Story 14.

STORY 14: A WET WEEKEND

Shift Manager Interviewee 7 recounted the story of a significant leak in a high pressure seal water system that made him seriously consider whether he should shut down that part of the plant immediately. The incident occurred on a weekend afternoon and the Shift Manager telephoned a maintenance supervisor to get his input. The supervisor gave a recommendation to dump the partly-processed batch of product and shut down the reactor straight away. The Shift Manager chose not to follow that advice immediately, but to take a few minutes to investigate other options.

If the water pressure in the seal system fell below the pressure in the reactor vessel, then the integrity of the seal would be lost and a serious leak of flammable and toxic materials to atmosphere could result. The Shift Manager chose to give one plant Production Technician the task of specifically monitoring the plant pressures that indicated the potential safety impact of the loss of water flow whilst the rest of the crew set about trying to put in place a temporary repair. 'I knew we were safe and I had the Control Room Production Technician monitoring the seal water pressure while I was outside so if we got to the point where we were even getting close to the seal water pressure getting the same as autoclave pressure then we would have had to [dump the batch], there was no question about that.'

In telling the story he volunteered that he felt that the leak needed to be stemmed within 30 minutes or so, or he would have needed to shut down the plant regardless of the stage of the batch at that point. The reason for this is simply that, whilst the plant was in this state, all the attention of the operating crew was directed to one highly unusual activity. They had no spare capacity to deal with any other plant issues that might arise. The Shift Manager was concerned about what could be called 'time at risk', although he did not use that description.

In this case, the water loss was stemmed within the timeframe that the Shift Manager had set himself and his crew. Production continued until that batch was fully processed and then that part of the plant was shut down to perform a permanent repair. 'I think the safety of my crew was the thing that was uppermost in my mind and I wanted to make sure that that was the thing we took care of before anything else, so I certainly wasn't about to put anyone in an unsafe situation ... We were all soaked by the end, but it was good. Not something that you would want to be doing every day but I think it was good that the guys were ... everyone wanted to make sure that we did the right thing and that we made it safe.'

Later discussion with the maintenance person who had been consulted revealed that his primary concern in recommending an immediate shut down was safety, but not specifically the potential for loss of containment. His concern was that running the system with insufficient seal water could cause major equipment damage that would require major in situ repairs by his crew that he knew to be complex and difficult (see Story 13). This would present the potential for further significant safety issues for maintenance personnel that the Maintenance Supervisor was keen to avoid. Since he gave his advice over the phone, he was happy to defer to the Shift Manager, who could see the extent of the leak and hence more realistically estimate the potential for stemming it.

This story shows many of the same considerations as those faced by nuclear power station personnel in similar circumstances, but also some important differences. The operating environment is different in that the physical plant and equipment has a much lower degree of redundancy than the nuclear power station. Also, chemical plant personnel have had no specific training in a process that encourages them to use their experience to fix limits, articulate them and stick to them. Despite there being no formal system in place, the approach of creating a situation-specific limit or rule was a practice that some people had informally adopted for themselves.

Reviewing the range of stories told, Shift Managers were less disciplined in sticking with their self-imposed limits than their counterparts at the nuclear power station. Several stories were told where the Shift Manager decided to continue operation on the basis that repairs would be put in place by a certain time, only for that time to be exceeded for other operational reasons. In two of the stories recounted in interview, this resulted in an adverse outcome and interviewees realized in hindsight that they perhaps should have stuck to their initial judgement. Story 15 is one such example.

STORY 15: FALSE ECONOMY

Shift Manager Interviewee 10 told the story of a faulty valve in the plant water system that developed late on a Monday afternoon. The fault meant that, instead of water levels being managed automatically, the Production Technicians had to manually check the water level in the system and ensure that sufficient water was in each tank. Maintenance advised that the necessary part would be delivered and installed on Tuesday. The water is used to cool each reactor vessel and so it is a key safety process system. The Shift Manager considered leaving the affected reactor offline until the valve was repaired, but was convinced without much difficulty by the departing day Shift Manager that they could manage to run the plant overnight by adjusting the water levels manually.

Interviewee 10 came back on shift on Tuesday evening to discover that the wrong part had been delivered during the day, so the system was still on line with levels being adjusted manually. He continued with this approach but '2 o'clock in the morning that decision bit us in the bum because the cold tank had dropped to a level where we lost suction to the pump that supplies the chiller units, which were no longer supplying chilled water to the tank, which meant that the autoclaves that needed chilled water weren't getting it and we had to [manually dump] two batches.'

This has both production and safety implications. Initiating a manual dump of the contents of the reactor indicates that the reaction is not under control and that urgent action is required. If the contents are not manually dumped and the pressure continues to rise, then the automatic relief system will initiate a dump to atmosphere, the 'last resort' in reaction control. Dumping the partially reacted product is also a production issue as that batch of product is lost and it takes some hours to get the plant back to a normal running state. 'So that's probably a really good example where in hindsight, the decision, what I said to [other senior operations staff] on Monday night, just leave it off until we get a new valve, would have been the best one.'

Story 16 is another interesting case. Despite the general feeling that safety decisions were quite uniform across the group, this shows that consultation does not always result in full agreement. It also shows another case where situation-specific limits used in initial decision-making may be allowed to slip.

STORY 16: JUST 'TIL TOMORROW

Interviewee 10 arrived at the interview late as he had been discussing a current operational problem with potential safety implications with Interviewees 1 and 4.

One of the chemicals used in the plant is stored in an atmospheric storage tank with an inert gas 'blanket' on the top to prevent contact with air. Following some maintenance work on the vent system approximately ten days before the interview, there were some problems with the inert gas system. Gas was flowing through the tank, but leaking out quickly so that the specified positive pressure could not be maintained and a low pressure alarm was showing continuously. Interviewee 10 explained that the need for the blanket was two-fold: quality (to prevent the chemical from decomposing and becoming less effective) and safety (the chemical is toxic and the gas blanket prevents build-up of toxic concentrations).

Interviewee 10 had been aware of the problem when it was first discovered, but he had been on days off since that time. He had arrived that morning to take up the position of on duty Shift Manager, and found to his annoyance that the problem had not yet been fixed. His first thought was to shut down and empty the tank with the problem. This would present a significant cost. The plant can continue to operate using an alternative chemical, but it is less effective and more expensive. Interviewees 1 and 4 had convinced him to wait for one more day, by which time the leak should be fixed. He said, 'One way or other they have convinced me that the new valve is going on tomorrow and that it should be fine until then. I've accepted that. I don't really like it.'

We talked about what he would do if the required repair were not done in the next 24 hours. 'Look they might convince me to leave it on but I wouldn't be happy about it. Because normally things like that come back and bite you.'

In other cases, if a safety system were offline (for example, for maintenance), then available alternatives would be put in place. Story 17 is an example of this.

STORY 17: TEMPORARY BACKUP SYSTEMS

Shift Manager Interviewee 2 told the story of an occasion where the Maintenance Department was doing some work on the fixed air monitoring system. This system is designed to detect leaks to atmosphere of the flammable chemicals used at the site and it covers much of the process area. On this occasion, a fault had developed with the system that Maintenance was unable to repair until the following day.

Interviewee 2 recounted how he decided, in consultation with the rest of the shift team and his manager, that they would continue to run the plant overnight without the fixed air monitoring system in place but with two temporary backups. The first was rigged up in the reactor area (which they judged as the most likely place for a leak to occur) using a portable analyser and some temporary tubing. Other parts of the plant would be covered by having one of the Production Technicians dedicated to circulating around the plant with a hand-held analyser, checking for leaks.

Operations continued overnight without incident and the fixed air monitoring system was repaired the following day.

These stories echo the attitude towards safety decision-making shown in other more general parts of the interview data. Shift Managers take a strong professional pride in their judgement and experience – in particular, their ability to produce plastic safely. Their judgement about safety relies not only on their technical knowledge of the plant, but in their real appreciation of the dangers associated with hazardous chemicals and industrial activities. In their view there is no conflict between safety and production, because safety always wins out.

4.3 Reporting and Recording Decisions and Incidents

Decisions made by shift managers were often reflected in operations records such as the shift log and keenly shared amongst the seven Shift Managers, but there were no other formal systems in place for specifically recording decisions taken. As at the nuclear power station, decisions made by the Shift Managers are invisible to the organization in any systematic sense.

This site operates an incident reporting and action tracking system similar in overall concept to the system at the nuclear power station. The events and stories recounted in interview as examples of decision-making are not normally recorded in that system. The exceptions are cases where (the potential for) some degree of loss was experienced. It was reported that the incident reporting system is for 'errors and faults' (Interviewee 2), not for unusual or interesting operational incidents.

4.4 Organizational Attitudes Towards Safety Decision-making

The concrete approach to safety and the reality of the hazards faced at the plant were emphasized in the safety induction training. The trainer (a maintenance engineer) focused on communicating the various safety rules applicable to a visitor, but the reason he gave as to why the rules should be followed was very different to the punitive compliance approach at the nuclear power station. The trainer emphasized the nature of hazardous chemicals on the site and the specific hazards posed by each one. He went on to describe past incidents at the site and with the same chemicals at other sites, and described in detail what would be seen if a leak were to occur. In common with the discussions in the later research interviews, the trainer spoke in very real and concrete terms of the dangers present and the ways in which they are controlled. He also mentioned a personal connection to someone who had been killed in an industrial accident (at another site). The purpose of this story seemed to be to emphasize that working safely is literally a life and death issue. The behaviour required of visitors to the site was explained in that context.

The chemical plant site operates under a business management system in which all work processes are proceduralized so that deviations can be identified and performance improved. The assumptions behind systems like this are discussed in some detail in Chapter 7, but the operational decisions made by the Shift Managers are curiously invisible in this system. In an attempt to proceduralize operational decision-making, the plant manager had introduced a formal process (called the Revolution Model). As discussed in Section 7.3, this system was ignored by the Shift Managers.

4.5 Summary

Shift Managers at the chemical plant were of the view that they make conservative decisions about safety issues at their plant based on good judgement and experience. They did not see rules or procedures as playing any significant role in decision-making about safety and operational issues. Some interviewees were adamant that neither process-related rules about how decisions should be made, nor goal-oriented rules about operating limits or boundaries would be of any benefit to them. This is despite many procedures being in place for specific issues such as maintenance work management (permit to work), incident reporting, behavioural safety observations. Although the use of these systems was not reviewed in detail, all interviewees implicitly acknowledged the importance of procedures/rules in these areas and gave the impression that these systems are comprehensive in scope and implementation. Shift Managers see them as very important and very useful.

Interviewees were generally proud of their professional problem solving abilities. Developing operational issues were seen as interesting and challenging puzzles to solve. One particularly satisfying result was to keep the plant running in the face of mounting operational difficulties for long enough to finish processing a batch of polymer. In dealing with specific problems, there were a number of cases described where interviewees had informally adopted the 'line in the sand' approach used at the nuclear power station. Unlike the power station, there was a tendency to let the self-imposed deadlines slip, sometimes with undesirable outcomes.

Interviewees were uncomfortable with the idea that safety and production can be conflicting goals. In the view of this group of managers, they never do things at their site that are unsafe. Interviewees articulated many personal stories about their experiences in explaining what constitutes safe and unsafe. The hazards associated with the plant were understood in a very real and personal way.

Chapter 5

Air Traffic Control – 'When you Kick a Ball you Don't Know Where it's Going to Land'

The third organization that participated in the research was Airservices Australia, Australia's air navigation service provider. Airline pilots regularly score close to the top in surveys seeking to identify the professions that hold the highest level of public trust.[1] Whilst air traffic controllers are not typically included in such surveys, they too are seen by the public as professionals who we are happy to entrust with our lives when we fly. As we will see, operational managers at Airservices feel that responsibility keenly.

5.1 Site and Operational Context

Research into operational decision-making at Airservices was carried out in the Melbourne Air Traffic Control Centre, the Sydney Operations Room and the Sydney Airport Control Tower. The staff in the Melbourne Centre are responsible for providing air navigation services in a large area stretching from the middle of the Indian Ocean, across central Australia and south to Antarctica. The Centre operates 24 hours per day. The number of controllers on duty varies according to the time of day (and hence traffic level), but is typically around 50. Unlike the power station and the chemical plant, the variable nature of the workload means that each controller has an individual roster and there are no fixed shift teams. Airservices employs approximately 1000 air traffic controllers in total across all control centres and regional facilities. The total workforce numbers approximately 3,000 people. In the Melbourne Operations Room, each controller sits at a console with radar displays and communications equipment to locate aircraft and talk to pilots. The controllers are divided into six rows and each group of two rows has an Operations Supervisor. Holders of this latter role have a desk literally in the aisle between the two rows of controllers they are supervising so they can hear (and often see) what each controller is doing.

Airspace is divided into volumes called sectors and individual air traffic controllers are licensed to work on particular sectors. Depending on the situation

---

1 Pilots came third in an Australian survey from 2011 described here http://www.readersdigest.com.au/australias-most-trusted-professions-2011/ and second in European data for 2011 here http://www.rdtrustedbrands.com/tables/community.shtml. Air traffic controllers were not included in either poll.

and location, air traffic controllers provide information and/or instructions to pilots about the route they are to take, weather conditions, other traffic in the area and the current location of the aircraft (if they are lost or deviating off course). The air traffic controllers deal with minute-by-minute operating problems within one sector or adjacent sectors, with occasional input from Operations Supervisors.

At the time the field work was carried out, the most senior operations personnel working in the Melbourne Operations Room were the System Supervisor (SS) and Operations Director (OD). These two positions are based at a large desk at the front of the Operations Room. Whilst they can see most of the activity going on in the room, if they are at their desk, they are too far away from the consoles to hear what individual controllers are doing. They have access to computer-generated information about issues such as traffic locations and weather and hence can see a technical overview of the entire region for which they are responsible (or any part of it) at any time. The Operations Supervisors and the SS report to the OD, who makes final operating decisions about issues that impact widely across sectors or have significant impacts on customers.

The SS is the primary co-ordination point between the air traffic controllers and maintenance staff. He reports equipment and software faults to maintenance and approves all work orders for routine maintenance and repairs. The overall system used by the controllers is highly reliable and the design incorporates a very high degree of redundancy. Despite this, there are occasional failures that impact the ability of some controllers to see the location of aircraft or other information (such as aircraft call signs) on their radar displays and/or speak to pilots. In the event of such a failure, the SS manages the process of repair and return to service in conjunction with the maintenance staff.

The OD and SS also deal with all external phone calls and visitors into the Operations Room and all calls out to stakeholders (airlines and other airspace users). This is a way of essentially managing the environment in which the controllers are working, both the physical environment of the room and the airspace environment as impacted by external factors (for example ad hoc requests for access). The OD and SS work closely together and essentially deputise for each other on a minute-by-minute basis since, depending on events, either of the roles may be temporarily overloaded. One of them is always formally in charge, as is made clear if either of them leaves the room.

The arrangement in the Sydney Operations Room is similar but on a smaller scale. The staff in the Operations Room are responsible for aviation traffic within 45 nautical miles of Sydney airport. A Traffic Manager supervises approximately six to twelve controllers (depending on the time of day). The Traffic Manager sits at a desk at the front of the room and essentially manages the environmental boundary for the air traffic control staff. Unlike Melbourne, several of the Traffic Managers hold current air traffic control licences and spend regular shifts 'plugged in', that is sitting at a console with a headset working as an air traffic controller. The Sydney Tower is staffed by seven positions (six air traffic controllers and a supervisor). Tower controllers manage aircraft arriving and departing from each operational runway, and traffic on the ground.

Airservices' system safety record is described each year in their annual report. The safety performance of the organization is also a key part of the broader view of safety in the aviation sector as a whole, as monitored by CASA and ATSB. For the last two years, Airservices have reported their performance based on a system of incident severity as shown in Table 5.1.

Airservices sets a target of zero incidents at SSI1 level. Their Annual Report (Airservices Australia 2007) states that there were four SSI1 incidents in each of the years 2004–05, 2005–06 and 2006–07. The organization also set and reported against targets for numbers of SSI2 and SSI3 incidents in various parts of their operating environment. Targets were met in some areas, but not in others. The Annual Report states that no 'pattern or systemic problem' was found in areas where targets were not achieved, although some specific remedial measures have been put in place.

Reviewing Airservices' safety record as part of the safety performance of the industry as a whole is also instructive. Following changes to the regulatory requirements for reporting of incidents in 2003, the ATSB undertook a major review of reported incidents in 2007 (Australian Transport Safety Bureau 2007c). The report focuses on incidents in regular public transport operations that is scheduled commercial flights rather than charter, private or sports aviation. The report found that, despite a 27 per cent increase in flights over the five year period studied, the number of safety incidents was either stable or declining. The

Table 5.1   Air traffic control safety severity incident classification (from Airservices Australia 2007)

Level of control resolved by (barriers)	Definition	Incident classification
Providence	Errors that either were not captured at all or were identified by airborne system defences such as ground proximity warning systems or traffic collision advisory systems.	SSI1
Pilot/other industry participants	ATC errors that were detected by pilots or other industry participants.	SSI2
Air Traffic Services, but not effectively	ATC errors that were identified and rectified by ATC but not in an effective or efficient manner.	SSI3
Air Traffic Services	Errors that were both identified and rectified in an appropriate manner by the ATS system.	SSI4

only exception to this is the number of breakdown of separation events.[2] In this category, the rate of incidents remains approximately constant but the number of incidents has increased (due to the significant increase in traffic). The ATSB analysis shows that approximately half the breakdown of separation events are caused by ATC procedural error and half are caused by aircrew errors. Of the events caused by controller error, two thirds were corrected by controllers and the remaining one third were corrected by the aircrew. As the ATSB point out, 'The final responsibility for maintaining separation between aircraft always falls to the pilots in command, irrespective of the services provided by ATC'.

Reviewing Australia's aviation safety record for more serious incidents (Australian Transport Safety Bureau 2007a), there were 14 accidents involving regular public transport aircraft from 2002 to 2006 inclusive. All accidents were investigated in detail by ATSB and there were no findings issued in relation to air traffic control as a result of these incidents.

The most significant findings by ATSB in relation to potential ATC causality of a serious incident in this period was for the fatal crash involving a private aircraft near Benalla, Victoria in 2004 (Australian Transport Safety Bureau 2006). In this incident the pilot and five passengers were killed. The aircraft was on a private flight from Bankstown to Benalla. The flight deviated 3.5 to 4 degrees from the expected track and triggered alerts to air traffic control from the Route Adherence Monitoring system. At the time, the aircraft was in a type of airspace where controllers are providing an advisory service, rather than directly controlling aircraft flight paths. Controllers knew that the pilot regularly flew from Bankstown to Benalla and they assumed that the deviation from the planned route was deliberate, so they did not alert the pilot. Weather conditions were such that the pilot would not have received any visual indication from terrain that he was not on course. The crash occurred when the pilot apparently began his descent towards what he thought would be Benalla, but was actually a heavily wooded area 34 km south-east of Benalla aerodrome.

ATSB found that, amongst other things, the occurrence 'demonstrated the need for effective communication between controllers and pilots to clarify any tracking anomalies'. Airservices conducted their own investigation, accepted the ATSB findings and made changes to ATC procedures and training aimed at preventing a recurrence (Airservices Australia 2006). Hopkins (2009b) has also studied practices at Airservices and published an account that described how other organizations could learn from them, particularly in the area of just culture and reporting.

Whilst the operational safety record of Airservices is far from perfect, it is a good performance. To date, there have been no major aviation accidents or incidents in which ATC error was a major causal factor. The increasing number of breakdown of separation incidents is of concern to the regulator and to Airservices

---

2   A breakdown of separation occurs when two aircraft are in closer proximity than the minimum specified in ICAO standards for the specific class of airspace.

and is an area receiving significant attention, particularly as aviation traffic is forecast to continue to increase.

5.2 Operational Decision-making

For both the chemical plant and the nuclear power station, if a significant operating safety issue arises, then the appropriate response is to shut the system down. Whilst in both cases there are hazards associated with the shut-down process itself, this is still the safest response to most serious safety issues.

At Airservices, the situation is different. Perhaps the most serious safety issue that can arise is an unexpected loss of a major part of the operating system. The physical system (equipment and software) that allows the controllers to locate, monitor and talk to the aircraft is highly integrated and includes multiple levels of redundancy. Air traffic controllers are trained in how to respond in the event of unexpected loss of communications or radar capability. There are well-established contingency plans and hierarchies to ensure that air traffic in any airspace immediately impacted by an equipment failure is managed to the ground (the safest place) in an orderly manner. This transition is relatively hazardous, so there are no cases in which supervisors would choose, for safety reasons, to immediately shut down the system for which they are responsible due to the potential breach of some kind of operating limit.[3] Having the system running, even in a degraded state, is always a safer option.

Despite this, the level of redundancy in the system must be reduced temporarily for short periods of time in order to allow for planned maintenance activities. The decision to release equipment for inspection, testing and maintenance lies with the supervisor. Story 18 is an example of this type of decision. Without a written rule set in place, this supervisor has adopted a conservative practice, similar to the nuclear power station example about cooling water pumps described in Story 9.

---

3   There may be some circumstances in which this decision may be made for security reasons. These types of contingencies were not discussed and are not part of the research described.

STORY 18: CALIBRATION AND TESTING

Runways at major airports are equipped with a radar navigation aid called an instrument landing system (ILS). This system allows appropriately equipped aircraft to land under conditions of reduced visibility since the radar allows the pilot and aircraft systems to 'see' the runway. Each ILS includes a backup power supply (in case mains power fails) which must be tested regularly to ensure that it is available should it be needed. Traffic Manager 4 cited this as something he would not allow if the relevant runway were in use and the weather conditions were such that instrument, rather than visual, landings were being made. His concern was that work on the backup power supply could result in unplanned failure of this operational system at a time when it is needed for safe landing of aircraft.

This decision is based on what each supervisor sees as good operational practice. There are no written rules on this subject.

On the other hand, there are also incidents that require immediate intervention from the air traffic control supervisors. These are in response to a significant, unplanned equipment or software system failure. As described above, the air traffic controllers have a standard set of contingency plans to manage aircraft in the air at the time of the system failure. Part of those contingency plans is to restore service as soon as possible and minimise schedule interruptions for aircraft. It is in managing the return to service that the supervisors are most involved in decision-making. Story 19 describes a case like this, although in this case the disruption was caused by external factors, not an Airservices system failure.

STORY 19: DELAYING ARRIVING AIRCRAFT

Operations Director 4 told the story of an incident in 2005 when a Thai Airways Airbus A340 landing at Melbourne airport blew a tyre and scraped along one runway through the intersection with the crossing runway (Australian Transport Safety Bureau 2007b). The immediate response to this potential emergency was managed in line with established rules (in the form of contingency plans).

There is another aspect of the required response that falls to the Operations Director and is not proceduralized. This is how to deal with further aircraft planning to fly to Melbourne. The incident had left both Melbourne runways temporarily unusable. Flights already in the air *en route* to Melbourne were diverted to other airports. In addition, Operations Director 4 decided to halt all departures heading for Melbourne that were scheduled to arrive within the 60 minutes following the incident. His logic was that this would give enough time to inspect the runways, start to clear the debris and establish the level of damage to the runway surface (if any). The trade-off he

---

was making was schedule disruption to the airlines versus the potential safety issue of having significant numbers of aircraft holding around Melbourne or diverting to other locations, depending upon how long it took to get one runway back into service. Note that he was guessing that a runway would be back in service before flights from destinations of greater than 60 minutes flying time (such as Perth and Brisbane) arrived.

His decision was unpopular with some airlines that would have preferred to have aircraft take off and hold at their destination, if necessary. This would allow them to land more quickly once the runway was back in service and reduce schedule disruption.

For less serious failures, there are safety and cost trade-offs regarding repair work. Limits are proceduralized in the form of System Restoration Times (SRTs). When equipment items fail, the urgency with which they must be repaired is specified by the SRT. SRTs have been developed based on the cost to repair balanced against the increased risk to aircraft and passengers of the item being out of service. Operational supervisors can request a variation to the documented times if they believe that particular circumstances warrant it. Story 20 describes a case such as this.

STORY 20: FIX IT NOW

Melbourne System Supervisor Interviewee 3 told the story of a day when the weather was very poor, with significantly reduced visibility, and the Instrument Landing System on one of the main runways at Perth airport failed.

The allowable System Restoration Time for this equipment is eight hours but operating without it in poor weather significantly reduces safety margins (and ultimately could lead to closure of the airport). The System Supervisor reported the failure to maintenance and requested the repair. He decided that the standard response was not adequate under the circumstances. He issued an exception report to maintenance requesting that the ILS be repaired as soon as possible. The service was restored in two hours. The operations group was very pleased with the service provided by the maintenance group and the exception report was later used to justify the extra maintenance costs.

Weather issues are another instigator of safety and traffic trade-offs. A set of fixed operating limits covers many potentially hazardous situations (for example maximum allowable cross wind on a runway), but these rules apply at a local or individual aircraft level. It is the job of the air traffic controllers, rather than the supervisors, to deal with these issues. In the event that a service interruption occurs in a specific location, supervisors generally get involved in the knock-on effects, making decisions about operating modes in other parts of the country due to, for example, significant weather delays at a major airport. These types of situations generally take some time to develop (and the longer the duration of the interruption, the more widespread the impact becomes).

The supervisors work hard to make sure they can manage such service interruptions proactively, rather than allowing the initial problem to compound into a major issue. Traffic Manager 4 described this as 'manoeuvring to keep the ship on course'. There are often direct trade-offs between safety and potential airline schedule interruptions in the form of rules applied by the air traffic controllers, but the supervisors try to prevent operations from getting into the situation where application of the rules is necessary. An example of this is that aircraft are not permitted to land if the cross-runway vector component of the wind is greater than 20 knots. This rule is applied by the tower controllers, but the Traffic Managers would hope to ensure that they consider the weather forecast in their selection of the duty runway(s) so that the tower controllers do not need to invoke this rule. Story 21 is an example of this concept in action.

STORY 21: THUNDER STORMS APPROACHING

After reviewing the weather forecast for the afternoon, when he came on duty at noon, Traffic Manager 2 reworked the aircraft arrival and departure slots for the afternoon peak in Sydney. This is done by a computer program with one of the inputs being the maximum allowable number of aircraft movements per hour. The earlier version of the afternoon's plan had been determined the previous day, using the maximum number of movements per hour based on requirements for noise sharing (see also Story 24). The forecast for bad weather, specifically thunder storms in the area around the airport, meant that, in the Traffic Manager's view, the maximum rate at which the controllers would be able to deal safely with arriving and departing aircraft would be reduced. The new maximum allowable number of movements per hour used to develop the plan was chosen based on the Traffic Manager's experience of similar situations. The new plan was then communicated electronically to the major airlines who must adjust their afternoon schedules accordingly.

The alternative course of action would be to leave the slots unchanged and for the controllers to deal with individual aircraft and their interaction with storms in the area as they arise. Thunderstorms impact traffic in three ways, as aircraft cannot safely fly directly through the storms. Avoiding storms potentially requires aircraft to follow non-standard approach and departure routes and, if the traffic levels are not reduced, storms create congestion as traffic has less overall airspace in which to fly. Also, local and rapid wind changes sometimes mean that runway changes are required at short notice to ensure safe conditions for landing. All of these aspects increase controller workload and hence the potential for error. As controllers reach the peak workload they can manage safely, flights are put into holding patterns and ultimately diverted.

This is the situation that the Traffic Manager was seeking to avoid by capping traffic movements in advance.

Due to the weather forecast the Traffic Manager also decided to:

* ask one staff member to start work one hour earlier in the late afternoon,
* defer some planned navigation equipment calibration work that would take one runway out of service and hence limit the flexibility of airport operations,
* request that the Australian Navy release some airspace over which they had previously been given temporary control for their own operational purposes.

Whilst these issues are all clearly within the Traffic Manager's area of responsibility, there are no written rules or guidelines covering this situation in particular.

On other occasions, unexpected bad weather can pose significant challenges, as illustrated in Story 22.

STORY 22: HAIL IN CANBERRA

A severe storm around midnight damaged communications equipment near Canberra so that two of the three radio frequencies used to communicate with aircraft in the Canberra area were not functioning. This meant that controllers could speak to aircraft aircrew as normal, but there was no further backup available. If there were any further failures of communications equipment, then controllers would be unable to speak to pilots of aircraft in the Canberra area.

Maintenance staff members in Canberra were called out to assess the level of damage and instigate repairs, but it soon became clear to System Supervisor 3 on night shift that the problems were not going to be fixed in time for the morning traffic peak. This was important as traffic heading to Sydney from Melbourne and other points south tends to be put into a holding pattern around Canberra if there are any delays. System Supervisor 3 (in conjunction with the relevant managers in Brisbane and Sydney) put in place two main contingency measures:

* An extra staff member with a detailed knowledge of Canberra traffic was called in to work in Melbourne Centre for the morning peak. This was 'insurance' in case communications to aircraft near Canberra failed completely.
* Sydney operations agreed to give priority to traffic from Melbourne, so that any holding required would be managed to the north of Sydney rather than in the Canberra area.

Traffic Manager 2 described his job as acting as a buffer for the air traffic controllers to ensure that there were enough staff, no distractions from phone or visitors, functioning equipment and that they were in the right operating mode for the prevailing conditions. Story 23 is an example of what he means.

STORY 23: TRAFFIC MANAGERS ARE A BUFFER

Traffic Manager 2 told the story of a recent foggy morning in Sydney as an example of his 'buffer' metaphor. On this particular morning, he spoke to Melbourne Centre and Brisbane Centre and told them not to release any departing aircraft bound for Sydney. Once the meteorology office staff were confident that the fog was clearing, he chose a time to start Sydney arrivals and worked back to begin departures from other cities to match that. In the past it would have been more common to let aircraft fly to Sydney and go into a holding pattern until fog cleared. This is a much higher controller workload.

Traffic Manager 2 emphasized that he made these decisions in consultation with supervisors in Melbourne and Brisbane. He felt this was a good example of the buffer idea.

Traffic Manager 4 said he believes that a key facet of experience is managing one's workload as a controller so as not to become overloaded. He thinks that inexperienced controllers have a tendency to take on too much and get themselves into strife. Part of the role of the Traffic Managers is to make sure that environmental factors do not encourage this. This would appear to be very similar to Traffic Manager 2's buffer concept. Melbourne Operations Supervisor 1 also made similar comments.

Another set of rules in the form of numerical limits applies at the supervisor level in some locations. These limits tend to be system performance-related targets or goals that operational managers should aim to meet, provided safety is not compromised. This is especially the case in Sydney, where airport operations are under some pressure due to noise in the surrounding residential areas from arriving and departing aircraft. Story 24 describes a case like this.

STORY 24: NOISE SHARING

I arrived to start workplace observations to find Traffic Managers 1, 3 and 4 in deep discussion about the best time to change from a north/south runway to the east/west runway, as required by noise sharing rules. Traffic Manager 1 was going off duty and Traffic Manager 4 was in the office primarily for non-operational duties. Traffic Manager 3 was on duty. The three of them had an animated discussion, with each person bringing up a range of factors that needed to be taken into account, such as:

- There were ten aircraft due to arrive within 20 minutes either side of the nominal changeover time.
- Changing runway would require aircraft to change their approach path, which would mean increased fuel costs, higher carbon emissions and increased controller workload.
- Several of the arriving aircraft just after the nominal changeover time were Boeing 747s (that is large aircraft), which would land on the longer north/south runway regardless of which runway was the nominated duty runway at the time.
- Some options would result in more work for the tower controllers as departing aircraft would have to taxi across the operational runway in some cases.

Traffic Managers 1 and 4 were keen to ensure that all factors were considered, but it was clear that everyone understood that it was duty Traffic Manager 3 who needed to decide when the runway change should occur. He spoke to the tower supervisor and the meteorology staff before deciding which flight would be the last one on the north/south runway. He then communicated his decision to the tower supervisor, the relevant controller in the Operations Room and the nearby Bankstown airport (which was impacted by his decision).

Similar decisions are required in some cases regarding controller workload. If staffing levels are suddenly found to be short (for example, due to someone calling in sick), the supervisor may be faced with a situation that requires a trade-off between cost (overtime), safety (increased workload for remaining controllers) and customer service (formally deciding that some services will be withdrawn). There are detailed procedures for how each option would be put into practice and also some firm constraints on the options available, but the decision regarding which operating mode is appropriate lies with the supervisor.

Undoubtedly, there are very many rules to follow. Traffic Manager 1 said, 'It's a very structured environment.' One supervisor described his thoughts about rules regarding air traffic control. He said that, despite the enormous number of rules and procedures under which the controllers and pilots operate, it was a myth that the outcomes were clearly determined by these rules (see Story 25).

**STORY 25: WHEN YOU KICK A BALL YOU DON'T KNOW
WHERE IT'S GOING TO LAND**

Sydney Tower Supervisor 1 said that he believed that most Airservices managers (and the general public) think that the entire job of air traffic control is fixed by rules. This is a myth. In fact the job is based on what he called guesswork. His explanation was 'when you kick a ball you don't know where it's going to land.'

He showed me an example unfolding in front of us. Arrivals and departures were alternating on the active runway. As we watched, the next departing aircraft approached the operational runway. The controller decided that it might be 'a bit tight' for the departing aircraft to be off the runway before the arriving aircraft reached the threshold point (the point at which it would be aerodynamically committed to the landing), so he advised the departing aircraft to hold. Whilst the separation distance between the arriving aircraft is fixed (and set up by the approach controller in the Sydney Operations Room), ensuring that only one aircraft is on (or committed to) the runway at any point in time is a complex judgement by the controller, based on aircraft speed and pilot behaviour.

In the view of Tower Supervisor 1, this is like kicking a ball, in that the controller uses a combination of rules and judgement to set in train a particular course of events, but once he has given his instructions (that is kicked the ball) the ultimate outcome is outside his direct control. His role is to manage the system so that it remains in a safe state, but the behaviour of the system is not always completely predictable or within his control.

In all of these cases, conservatism was seen as the key. Traffic Manager 6 said: 'The key to safety in this job is to be conservative. It is better to overreact in that situation on the off chance that things might go bad than to assume everything will be OK and get caught out.' He has been in the job for 30 years. Story 26 records a memorable event for another supervisor that, to him, emphasized the need for conservatism.

STORY 26: THIRD TIME UNLUCKY

Operations Director 5 also emphasized the need for conservatism. He told the story of something that had happened to him many years ago when he was an air traffic controller in Launceston (a small regional airport).

A light aircraft was flying through the section of airspace under his control. The aircraft was registered as flying under Instrument Flight Rules, which requires the pilot to call in to the controller at specified intervals. The pilot failed to report in, so the controller called him to remind him. The pilot said 'oh yes, sorry, I forgot.' At the next reporting point, the pilot again failed to call and so the controller called him again. Again the pilot apologized.

Some time later, the aircraft again failed to report in. The controller considered just ignoring the situation, but decided he should chase the pilot for a third time. There was no response. The aircraft had flown into a hill and needed a full emergency call out. Operations Director 5 said that at the time he felt a bit like 'the boy who cried wolf', but that this experience taught him something that he has never forgotten, which is that he must always assume the worst has happened until proven otherwise.

Anticipation, conservatism, buffering are all key factors in the decision-making of the Operations Directors, System Supervisors and Traffic Managers. At their level, operational decisions are mainly based on experience and judgement rather than rules. In the few cases where rules exist for decisions at operational management level, outcomes are still uncertain as the title of this chapter suggests. Supervisors can actively manage the system, but there is a lag time before their actions have an impact, and the effect of changes that they instigate is not always completely predictable.

5.3 Reporting and Recording Decisions and Incidents

Incidents, events or failures are reported, recorded and managed at Airservices using three separate systems. Firstly, equipment faults and failures are reported and recorded in a system that is aimed primarily at management of repairs. This system is the primary written communication link between the operations and maintenance parts of the organization and hence is designed primarily with work management in mind. There is a broad link to organizational learning in that data from this system is also used for trend analysis about the overall reliability of the system. This is of little direct relevance to the operational managers.

Secondly, Airservices operates a comprehensive system for reporting of operational incidents that occur within the broader airways system. Airservices staff are in the best position to identify incidents, given the nature of the view they hold of the system as a whole. The system manages several thousand reports per year (Hopkins (2009a) reports 135 reports in a sample week). Approximately 10 per cent of incidents have a causal component related to air traffic control. Whilst anyone is free to report an incident in this system, the Operations Directors, System Supervisors and Traffic Managers make many of the reports for incidents that occur during their shifts, as the controllers themselves are often too busy. Such reports were made either as a result of the manager's direct experience or on behalf of controllers. Examples observed included:

- A flight from an international carrier that flew an unusual route across the continent.
- A breakdown of co-ordination between one controller and the controller for the adjacent sector that was reported by the controller who made the error.
- A light aircraft that appeared on radar but could not be contacted on the standard frequency. This report would also go to CASA, who would write to the registered owner of the aircraft.
- Several cases in which a light aircraft ventured into the wrong area in a violation of controlled airspace.
- A light aircraft pilot who became lost and had to be assisted to make an emergency landing.

All reports raised in Melbourne go to the Operations Directors by email. In Sydney, Traffic Managers receive a summary email from one of the local safety specialists who screens all the reports for items he believes will be of interest to them. Details of specific incidents and also overall operational trends were shared and discussed in a variety of ways.

Airservices has an important role in improving operations for the entire system due to their unique view of the airways system as a whole. One small example of this came to light at a safety meeting observed in Sydney where one of the tower staff had extracted data about one type of incident being increasingly reported. There had been a number of recent incidents in which aircraft arriving in Sydney had had to 'go around'. The air traffic controllers set up arriving and departing aircraft so that they are spaced or separated safely, but high traffic levels lead controllers to operate close to the specified standard for the minimum separation. For arriving aircraft, there is a point of no return called the threshold point. If an arriving aircraft reaches that point and any departing aircraft have not completely cleared the runway, then the pilot must abort the landing and 'go around' to avoid becoming aerodynamically committed to landing on a runway that may not be free of obstacles. 'Going around' causes schedule disruption and uses extra fuel and so is very unpopular with airlines. From the perspective of an individual pilot, it may seem like an error on the part of the controller that the arriving and departing aircraft are not appropriately spaced. In fact, it is the result of such factors as the dynamic interaction between the speed of the arriving and departing aircraft, and the behaviour of pilots in clearing the runway in addition to the actions of the controllers. This is the basis of Story 25 and the title of this chapter.

A staff member had completed a quantitative and qualitative analysis of the information and put forward a possible explanation as to why several of these incidents had occurred recently in Sydney. The discussions in the safety meeting centred on the need to talk to the airlines about apparent changes in company policy on aircraft arrival speeds. This issue appeared to be contributing to the increased number of reports as aircraft appeared to be arriving at the threshold point earlier than controllers were expecting based on their previous experience, and hence sometimes before the departing aircraft had cleared the runway.

Another example of this aspect of Airservices' role is their program to reduce violations of controlled airspace. These are cases where small aircraft have inadvertently strayed into areas that they are not registered to use, thereby presenting a potential hazard to larger jet aircraft. When air traffic controllers notice such events they are reported, and CASA writes to the aircraft owners to advise them that their behaviour is not appropriate. Airservices has a campaign to investigate the reasons behind such incidents, including discussions at regular industry liaison meetings and also an on-line survey hosted on the Airservices web site[4] where people who receive a CASA notification are invited to report the reasons for their actions.

The third type of incident reporting is a confidential system whereby any Airservices staff can report concerns directly to the corporate safety management group. Forms are displayed in several locations at each operating centre, but none of the operational staff mentioned having used this system.

These systems all play a significant role in safety and are well-integrated into the daily activities of the Operations Directors, System Supervisors and Traffic Managers, but, similar to the other organizations studied, the incidents of most interest to these operational managers are not reported into these systems. Such occurrences are shared informally at shift handover and in other informal one-to-one conversations.

5.4 Organizational Attitudes Towards Safety Decision-making

Airservices operates under a regulatory regime based in international standards and rules for air traffic control (as defined by the International Civil Aviation Organization – ICAO). ICAO requirements are detailed and onerous, and Airservices management and staff take their responsibility to meet those requirements very seriously. This provides the basis, and much of the detail, for the rules that cover air traffic control operations and thus attitudes towards safety by senior management tend to be heavily focused on rules and compliance. One senior safety manager told me that this research would show little other than compliance with rules, as all safety decisions are specified in the various procedures and nothing that is important for safety is left to judgement.

The belief that everything of safety importance is proceduralized leads to the conclusion that those things that are left to the discretion of the operational staff are not safety-critical. Further, this means that senior management feel at liberty to question the judgement of managers over safety-related items. Airservices is the only one of the organizations studied in which operational managers felt under pressure to change their decision-making practices to be less conservative and hence, in their view, less safe.

---

4   See http://www.airservicesaustralia.com/vcasurvey/default.asp accessed on 10 April, 2008.

This is a direct result of a desire to increase operating efficiency and reduce costs. The organizational view is that, for those items not specified by (ICAO) rules, safety becomes a matter of risk management and cost savings can be achieved by changing practices in areas that are overly conservative.

This is discussed at length in later chapters.

5.5 Summary

The outstanding theme related to safety decision-making by operational managers in an air traffic control environment is anticipation.

The safety decision-making focus in the other two organizations is on identifying if or when it is necessary to interrupt or stop operations at the site in the interests of safety. In contrast, at Airservices it is *never* the safest option to shut down their own operations without notice. Continuing to run the air traffic control system, no matter how degraded, is always safer than operating without it.

The effort of the operational managers goes into managing maintenance and operational conditions to ensure that the air traffic control system (equipment and people) is never compromised. They take professional pride in anticipating potential problems before they develop and making changes to overall operating parameters to ensure operations proceed smoothly. In a sense, they are operating at their best when superficially it could appear that nothing has happened, because frantic activity means that they have failed to adequately anticipate a problem.

This creates an organizational dilemma in an environment of cost reduction. The adjustments to operations to prevent anticipated problems are not generally proceduralized but are based on the experience and judgement of the operational managers. They often involve additional costs (typically overtime) and these are not popular with more senior management. Looking at the operating record historically, nothing has happened, and the assessment of whether these decisions are appropriate or wasteful becomes one of perception – a question of what might have been. Similar to the attitude at the nuclear power station, the operational managers were unanimous that conservatism was always the best course of action.

On the rare occasions when the air traffic control system (partially) fails or some external factor such as unexpected bad weather occurs, there are more obvious problems for the operational managers to deal with. In these cases, their focus is still on anticipating further problems, and on recovery (as the immediate operational responses to such situations are proceduralized and dealt with by individual controllers).

The operational managers at the two locations studied have a strong loyalty to their group of peers and a strong sense of responsibility to the flying public. They are much more cynical about the priorities and directions of their employer organization.

PART B
Acting Both as Employees and as Professionals

The preceding three chapters described in some detail the decision-making practices in three high hazard organizations. The next five chapters look more closely at key themes that emerge from these case studies and how they impact decision-making.

The operational managers in the three organizations work in very different technical environments, but there are many common features in the ways in which they describe their experience of making safety/production tradeoffs – the types of stories they tell, the relationships with others in their organizations (peers, other members of the operating team and organizational superiors) and what they feel is important in their decision-making. As would be expected, the technical aspects of the stories vary widely but the stories themselves have many similarities.

Firstly, operational managers across these three organizations saw their role and their responsibility for safe operations in a very similar way. Issues of occupational role and identity are discussed in Chapter 6. The distinction between seeing oneself as an employee or as a professional is explored, along with the implications for safety of these dual occupational identities.

Another key theme in the research data from all three organizations was the use and role of rules - written procedures governing specific processes and/or defining fixed operating limits. All three organizations were grappling in various ways with the tension between compliance with written rules on the one hand, and the role of experience and expertise on the other. Chapter 7 addresses rules and compliance, including the influence of quality management principles on safety decision-making.

Professionalism is associated with qualities such as loyalty within the profession, a sense of vocation, identification with the goals and values of the profession, integrity and public trust. Chapter 8 reviews the impact of these professional characteristics on decision-making by operational managers. The discussion focuses particularly on stories. Interviewees seemed to have little trouble recalling and recounting specific episodes describing past decisions. These stories initially led to broader discussions about what it means to be safe or unsafe and other relevant issues, which then often led to more stories about past experiences. The role of stories is also discussed further in Chapter 8.

Decision-making seen as sensemaking is a process with an important social dimension. Professional relationships with peers, other members of the operating team and organizational superiors are discussed in Chapter 9. These relationships are important in making the best decisions in the face of uncertainty in complex systems.

Operational managers were adamant that they do not operate unsafely – that there is no conflict between safety and production. This seems surprising in an environment where risk assessment (which balances safety and cost) is such a common way of viewing safety management. These decision-makers do not consciously experience what Hollnagel (2009) calls the efficiency-thoroughness trade-off or ETTO. Rather than providing an alternative generalized explanation, most operational managers told one or more stories, as described earlier. This range of stories provides an opportunity to seek common themes and explanations as to how the operational managers consider multiple organizational goals such as production and safety - and how they make sense of abnormal operating situations. This is explored further in Chapter 10.

Chapter 6
Decision-making and Identity

Operational managers are, on the one hand, employees. They have a place in the organizational hierarchy with defined roles and responsibilities. They must act in accordance with organizational goals as communicated by their organizational superiors. On the other hand, the research data shows that they have a second identity linked to their work – as professionals in their given field. On most occasions, the actions required of them both as employees and as professionals are the same. This should come as no surprise. After all, organizations are generally keen to avoid accidents and all operational managers see maintaining safe operations as an important professional goal, too.

These dual identities as employee and professional mirror to some extent a contrast between bureaucratic organizing and professional organizing. This contrast has a long history in social science (particularly sociology) but has been little explored in the context of modern industrial safety practices. The key issue for safety performance is the reliance unwittingly placed by organizations on professional qualities of their operational managers such as experience, technical knowledge and a concern for public trust.

6.1 Organizational Identity

Before looking in detail at how operational managers make specific decisions, it is appropriate to consider some broader questions about the role of operational managers in ensuring safe operations. What do operational managers believe are their responsibilities for safety and production? How do they conceptualize their role in relation to overall safety assurance?

The starting point for this discussion is occupational identity. All operational managers interviewed for this research were long-term members of their employer organizations and also very experienced in their chosen professions. This gives them two strong occupational identities – as an employee and as a member of a professional peer group. These senses of self or aspects of their identity each play a key role in their safety decision-making. When acting from his identity as an employee, an operational manager works within the framework of company goals and priorities. In coming to his decision on the way forward, he will be inclined to follow company expectations on how decisions are to be made – following rules and procedures as laid down. On the other hand, when acting from his professional identity, an operational manager acts from a different set of values and priorities forged as a member of a tight-knit and professional operating team.

Professionalism invokes such concepts as vocation, public trust and authority deriving from knowledge, rather than organizational position. Middlehurst (1997) defines professionals as having the following features:

- Technical and theoretical expertise and the authority and status flowing from such expert and highly valued knowledge, understanding and skill;
- The establishment and the exercise of trust as a basis for professional relationships (with clients and between professionals);
- Adherence to particular standards and professional ethics often, but not always, represented by the granting of a licence;
- Independence, autonomy and discretion;
- Specific attitudes towards work, clients and peers involving dedication, reliability, flexibility and creativity in relation to the 'unknown'.

Middlehurst described the importance of trust relationships between professionals and their clients and of professional ethics. Other authors (Friedson 2001, Sullivan 2005) expand on these ideas to describe the strong sense of responsibility held by professionals for the public good. Professions have strict entry standards in the form of long training in both theoretical and practical considerations and often licensing arrangements. This training and induction into the culture of the profession engenders members of this exclusive group with loyalty to their peers, rather than to their employers.

Identity has a direct bearing on decision-making. Many people and many organizations believe that good decisions are made on the basis of rational choice so that when faced with a problem, the decision maker identifies options available, analyses them and chooses the best one based on rational, generalized criteria. Researchers from the field known as naturalistic decision making or NDM have studied how fire fighters, intensive care nurses and other highly skilled specialists make time pressured high stakes decisions. In contrast to the idea of rational choice, NDM researchers have shown that decisions like these are made based on (often unconscious) pattern recognition. Weick (1995) takes these ideas further with his work on sensemaking. In this way of viewing decision-making, actions are a natural progression in the moment of the sense that an individual makes of changing circumstances and events as they occur. This leads to the idea that decision-making is not a rational process occurring in isolation of context. Instead, decision-making is seen as a richly context-dependent process. Sensemaking emphasizes that the sense we make of the world around us literally depends on our sense of identity or role. Weick (2005) describes identity as 'who we think we are', but he acknowledges that our identity is not singular, nor is it developed in isolation of our environment. Our behaviour provokes a response in others which reinforces or weakens our identity. Each of us has a range of ways in which we see ourselves, perhaps an 'electorate of selves' who in each moment choose which identity is best to prevail. There is significant commonality in the view of safety and production issues from these two perspectives, but they are not always identical, as is described in the following sections.

Insights into the organizational expectations of operational managers can be found in formal job descriptions and the authority that comes from those descriptions for safety and production-related decision-making. As described in Section 6.2, the role of the operational managers is primarily defined in professional, rather than managerial, terms. Operational managers hold absolute authority for continuing operations on a minute-by-minute basis. This links their responsibilities for both safety and production, but they are largely insulated from management responsibilities for finance and cost control. This gives them a high degree of authority, based on their extensive operational knowledge.

Section 6.3 explores a number of cases where the behaviour of the operational managers suggests that they sometimes act from their professional priorities, rather than in accordance with stated organizational goals. Operational managers can sometimes put themselves under pressure as a result of the way they see their primary professional task – producing electricity, making plastic or moving aircraft. This unacknowledged source of pressure to produce, related to professional identity, has clear safety implications.

Another identity issue considered in this chapter (see Section 6.4) is experience, in the sense of length of service, and the perspective this gives operational managers about safety and production issues. The stability of employment of the operational managers is a reflection of their individual sense of vocation and gives them a multi-layered view of the system and how it operates.

A sense of integrity and public trust are other professional qualities that feature in the research data as described in Section 6.5. The work of the operational managers at the air navigation service provider has the most direct link to members of the public. They show a high degree of dedication to public service in their day-to-day activities. Given the nature of the technologies involved, all operational managers have a major role in maintaining public safety. They take this very seriously at all sites, adopting an attitude of protector.

As summarized in Section 6.6, each of these aspects of occupational identity has implications for safety decision-making and hence overall safety performance.

6.2 Organizational Authority

In all three organizations, the ultimate responsibility for deciding that the technological system should stay on line lies with members of the operating team. This is normal for such facilities where delays in decisions to intervene in production have contributed to past accidents and was a lesson well learned after the Piper Alpha disaster in the 1980s. In that incident, 167 people lost their lives when an offshore oil and gas platform in the North Sea was destroyed by a series of large fires. The platform was a hub for a series of collection pipelines from other platforms in the area. One factor in limiting the inventory of fuel available to burn was the time taken by the managers of adjacent facilities (who could see the fire) to stop production of oil and gas. In at least one case, there was a delay in shutting down until the facility manager received permission to stop production from more senior management in the onshore office (Cullen 1990).

So, in one important area, the operational managers hold ultimate organizational authority. They are given this authority because of their professional expertise and yet this level of authority does not apply to everything they do. The positions held by the operational managers are all in upper middle management, two to four levels below the CEO of their respective organizations so, as employees, they hold senior positions but there are limits to their authority. There is one major exception. At the nuclear power station, the authority held by the Shift Managers is formally documented as part of their position description and also emphasized in the Conduct of Operations Manual which states: 'If management representatives are present during operations there should be no doubt who is responsible for decision-making'.

The writer is trying to say that, even if senior management representatives are in the control room, authority for making operational decisions still lies with the operating crew, although oddly the form of words chosen could be interpreted to mean the reverse. At the air navigation service provider, this concept is known as Operational Command Authority and is respected in practice to the point that, if the operational manager holding that authority is to leave the Operations Room (even if only for a few minutes), he verbally hands over that authority to someone else before departing. Similarly, at the chemical plant, where the arrangements were not so formal, everyone was clear that the ultimate decision about the continued operation of the system was the responsibility not of site management, but of operations staff. In practice, this was often the on-duty operational manager, although some aspects of operational authority may lie with other members of the operating crew.

Final operational authority lies with the professional operating crew in all cases, even when their organizational superiors are present, because they are trusted as having the ability to make better operating decisions than any other person. Becoming operational manager in any of these organizations can be seen as reaching the highest levels of a 'career of achievement' (Zabusky and Barley 1996) – where seniority is based on professional skills and knowledge. This is also clear in written position descriptions, for example the published document detailing role, responsibilities and accountabilities for air traffic control supervisors makes it clear that these positions are technical operational positions. The prime responsibilities relate to safe and orderly delivery of services, in particular exercising Operational Command Authority. As written, these positions have no responsibility or authority related to costs, budgets or financial performance.

Despite this straightforward description, operational managers were subject to some significant conflicting pressures, related to the co-ordinating role of the organization as a whole. The air navigation service provider manages the day-to-day operation of Australian airspace on behalf of the government. This involves balancing safety and environmental aspects with the needs of the various airspace users (international and domestic commercial airlines, private commercial users, specialist airspace users and sports aviators). This means that in the air traffic control situation, operational managers are also working with a range of conflicting goals. Uniquely in the organizations studied, this includes direct pressures both from within the organization and externally. In any given situation, the required trade-off typically involves three parameters – safety, cost to the organization and customer service. (In the case of Sydney airport operations, there is a fourth socio-political goal which requires regular changes to the duty runways to ensure that aircraft noise is shared equitably amongst residents in various parts of Sydney.) It may come as a surprise that safety of passengers and customer service can be competing priorities. Any airline would surely claim that passenger safety is its number one priority. Nevertheless, air traffic control staff have a system-wide view that is not available to individual pilots and, as a result, sometimes make decisions in the interests of the safety of the system as a whole. Such decisions can cause schedule delays for individual aircraft and hence are not popular with those pilots impacted. Decisions requiring an operational manager to balance all three issues are common; for example, an unexpected shortage of rostered staff, which can be solved either by bringing in extra people to work overtime or making a small reduction in the level of service provided. The most appropriate choice depends on the safety implications of the specific weather conditions, location and expected traffic level, and the ATC supervisors had the ultimate say over operational decisions and no direct budget accountability.

On the other hand, the Shift Managers at the nuclear power station were specifically required to address overall system efficiency (including cost) as well as safety production. According to the company documentation, the purpose of the position of Shift Manager is: 'To manage the resources at his disposal to ensure that the power station is operated in accordance with Statutory regulations, Company regulations and procedures to meet the agreed programme. Within these constraints, to achieve the optimum operating efficiency commensurate with the need for plant security, availability and safety.' On the other hand one Shift Manager said:

> I think we are very lucky in this industry in that we have not got shareholders or the perceived shareholder effect where profits come first. I've no objections at all to say 'shut both bloody reactors down – I'm not happy'. I don't worry about the cost. I worry more about how much effort it takes to get the bloody things back, but with regards cost, I don't worry.

The role of the operational managers at the chemical plant was defined in a similar way but despite this general requirement to balance multiple goals, at both the chemical plant and the nuclear power station, operational managers were insulated to some extent from the direct cost impact of their decisions. At the chemical plant, the financial performance of the company can be impacted by operational decisions in several ways including variations in both feedstock costs (for example shipping demurrage) and product delivery (for example contractual conditions regarding timing and quality). Operational managers were certainly aware of these aspects of the organization's financial performance and were required to take them into account in general terms, but they had no direct responsibility or accountability in these areas. Feedstock contracting and product sales were handled by other departments separate from Operations. Operational managers and their crew never spoke to suppliers or customers and hence were not subject to direct pressures from those sources. They also were not held accountable for budgets or contract financial performance. In a similar way, whilst they were aware of the general need to keep costs under control, they had no direct responsibility for staff costs, including things such as overtime budgets, so the decision to call out maintenance staff outside normal working hours (for example) did not result in any direct financial feedback to the operational managers. This was explained by senior management as deliberate policy on the part of the organization to reduce possible financial pressure on operational decision-makers.

The situation is similar at the nuclear power station. It was described how the organization has moved to a more commercial focus in recent years so that generation output has come to be equated with income (or contractual penalties for power supply interruptions). Again, whilst supervisors were aware of a general need to maximize income and keep costs under control, they had no direct responsibility for meeting financial or budgetary targets. They also had no professional dealings with customers, except at an operational level.

In one of these organizations a significant organizational change was in the early stages of implementation. Operational supervisory positions were being reframed as management roles, rather than technical positions, and the new job descriptions specifically included responsibility for staffing budgets and other costs. Holders of the new role would not be permitted to hold operational licences and hence would be unable to participate directly in operating the system. The prime focus of the revised role was leadership and management of a team to deliver safe, efficient and cost effective business outcomes. The new position description stated: 'On balance, the substantial work of the role is to provide sound managerial leadership to achieve the required outcomes of the team' and 'The [manager] ensures that their own work and the work of the team is integrated with other shift teams and aligned with business objectives'.

This represents a major change to the nature of the operational manager positions. Holding this position under the original description held a significant degree of professional prestige and bestowed status based on professional reputation – the pinnacle of a 'career of achievement' (Zabusky and Barley 1996), as described earlier. The revised description de-emphasizes professional skills and emphasizes position in the organizational hierarchy. Being appointed to such a position is a step on a 'career of advancement'. One senior manager described this as a deliberate distancing of the position from what he believed was the over-conservative (and hence costly) view of the operations personnel as to how things should be done. Another senior manager indicated that the redefined job role was partly in response to staff feedback from younger employees who had expressed some frustration in employee satisfaction surveys about the lack of career path available. On the other hand, those who currently held what were effectively senior technical roles were mostly unimpressed by the changes. Several expressed frustration or anger at the way the new role was defined, seeing this as an attempt to introduce a major and unwelcome change in their professional priorities.

In summary, review of the formal role definitions of the operational managers shows that, in all three organizations, the roles merge typical responsibilities of employees (such as obeying company standards and meeting organizational goals) with professional characteristics (such as independence and autonomy). Whilst not always emphasized in role definitions, organizations also give significant authority to operational managers in the sense of having complete authority for minute-by-minute operational decisions, including ultimate authority to stop production for safety or other reasons. Whilst this gives them significant organizational authority, they are still required to follow policies and procedures for many facets of their role, including operational decision-making. In this context, authority lies with their organizational superiors, and operational managers are expected to adopt organizational goals about safety and production as their own, and to follow decision-making procedures laid down for them.

Looking at where final accountability for production costs lies also gives clues as to how organizations perceive the role of the operational managers. All three organizations had independently developed structures that separate the specific accountability for internal costs from safety/production decisions. In addition, whilst production targets are well understood, generally speaking those making safety/production trade-offs are insulated significantly from the direct pressure of clients, customers and downstream users. The reorganization plans in one organization are an exception to this model and are driven at least partly by the requirement to move operational decision-making away from professional judgements and closer to the goals of the wider organization.

6.3 Professional Priorities versus Organizational Priorities

An ability to effectively manage the tension between safety and other goals such as production is part of the definition of an HRO. Perin (2005) calls this tension 'the infrastructure of conundrums'. The previous section highlights how some organizations manage the pressure on operational managers to keep the system running in order to meet cost or efficiency goals. But there are other reasons why a senior professional may choose to keep the system on line such as professional pride. The broader attitude of the operational managers towards production gives an insight into other reasons they may favour production over safety.

At the nuclear power station, the subject matter covered in interviews ranged widely, but operational managers omitted any mention of one key organizational issue: that the power station was shortly to be permanently shut down. One nuclear power station manager described his profession explicitly as being about generating electricity. Another operational manager said his job was to 'keep the lights on' in other words to ensure continued electricity supply to the national grid. Yet another spoke of the relativities between the nuclear power station and the nearby wind farm in terms of the end users that each generation source could supply, thereby emphasizing the link between his job and the public service he helps to provide. The truth was, however, the power station was very close to the end of its operational life and several hundred people were working at the site on plans and projects to commence decommissioning only a few years hence. This is far from a trivial exercise at a nuclear power station and the plan extended for more than 20 years.

Most of the people working on decommissioning plans were new to the organization. They were physically located in a series of temporary huts behind the main administration building, well separated from the operating team. In addition, they reported to the Business Manager, who reported directly to the Site Manager. The Business Manager was a newcomer to the site and did not have a good relationship with the rest of the management team, many of whom had worked together for an extended period. The relevance of this to discussion about the professionalism of the operational managers is that, despite the high profile of decommissioning activities going on at the site in terms of management attention and numbers of people, not a single one of the operational managers mentioned the upcoming change in the status of the site. It was as if, for them, this large part of the organization and its plans did not exist. This significant near-term organizational goal remained unmentioned.

It could be said that the decommissioning plans were not directly relevant to the subject of the interviews and hence they were not discussed. In fact, the interviews were conducted in the evening (as that is a quiet time on site) and many of them extended in duration and scope far beyond the specifics of the stories of operational decision-making that were the core subject. Several of the interviewees had come to the site from another nuclear power station that had itself been shut down only a few years earlier. In this context, it was very odd that no-one mentioned that this site would shortly be suffering a similar fate. Operational managers communicated a clear pride in their (individual and collective) knowledge and skills, and their professional achievements. There was an odd dissonance in that these conversations were taking place in a facility where the organizational goal was for operations to cease.

Similarly, at the chemical plant one significant aspect of current operations where professional objectives and those of the organization did not coincide again related to an organizational need to limit production. In this case, almost every interviewee raised this issue and many complained long and loud about it. Operational managers at the chemical plant had been forced by commercial circumstances into an unusual operating mode that was universally detested. For reasons related to the worldwide market and demand for feedstock in China, at the time the research was done, the chemical plant did not have sufficient contracted feedstock supply to run continuously. Instead of running the plant for most of the year and shutting for a few weeks to perform annual maintenance (the norm for plants of this type), the plant was operating in short runs of several weeks interrupted by shutdowns of a few days to wait for new feed material. Work required to be done at the annual shutdown was distributed into the smaller windows. This way of operating was the most attractive commercial option available for the company in the current circumstances, but it was universally hated by the operational managers. Several of them claimed that the plant itself hated it too. For the maintenance crew, this operating mode was certainly hard work with lots of planning sometimes to no useful purpose (as sometimes, for commercial reasons, the planned short shutdowns did not take place), but the operations staff had the most vocal, almost visceral, reaction. This operating mode seemed to be preventing them from doing what they felt was their prime professional goal – making good quality product. Their professional identity reacted strongly against the need to constantly interrupt production, despite the wishes of their employer.

This general attitude of favouring production clearly has potential safety implications when responding to specific situations. It is a matter of the operational managers' professional pride. They come to work each day to produce electricity, produce plastic or move aircraft. Allowing a given situation to get the better of them so that they need to stop production was seen as failure in professional rather than organizational terms. One of the nuclear power station staff captured the tone of his colleagues' conversations well when he said: 'There is still a pressure. There is still an internal one. Our internal pressure to get the plant on. That's our profession, that's what we're about ... but we do know that if there's an issue and we decide "no we're stopping" we don't have to spend all day justifying it'.

At the chemical plant, the strong professional preference of the operational managers was to see the plant through to the end of a batch. If things went wrong, they had a strong desire to get to the end of the batch before shutting the plant down (for example, see Story 14). There was a similar sense here, too, that shutting down before the end of the batch meant that the plant had somehow got the better of them.

At the time the research fieldwork was carried out, a major operational restructure was underway at the air navigation service provider and many existing operating practices were being challenged by the organization – a situation that was creating a conflict between professional and organizational priorities for operational managers. Senior management were asking operational managers to make changes to some long-established practices that management felt were overly conservative and unnecessarily costly, although, at least with regard to operational decision-making, it was emphasized that the final decision (and responsibility) was always with the operational managers. The responses of the operational managers varied. Some felt that their professionalism was being challenged. They stuck with past practices and responded with hostility to the requirement to acknowledge that there was a cost linked to their decision and hence they should be able to justify it. Others moved to the new way of operating and, when asked why they were happy to change when some others were not, responded that it was clear what the organization wanted them to do, so as employees they felt it was their duty to do what was required of them. This was the clearest case in the research data where professional and organizational identities seemed to be promoting different behaviours.

The complicating factor in this particular case is that many of the controversial issues related to staffing arrangements (such as the need for overtime), so it was easy for senior management to see the resistance to change as being at least partly driven by self-interest. It is not possible to judge here the technical pros and cons of the disputed practices, and to a large extent that is irrelevant to the issue at hand, but the safety of the system as a whole relies on the professionalism of the operational managers and they felt that their professional judgement was under attack.

As these examples make clear, operational managers are influenced in their overall disposition towards production, and their specific decision-making, by their professional identity and professional goals, as well as those of their employers. Whilst these goals often coincide (at least with regard to safety), this is not universally the case. Efforts to foster good decision-making should therefore take this issue into account.

6.4 Experience

Since identity is based on interaction with others, another significant factor in developing occupational identity is in the length of time that the operational managers have been part of their professions and part of their employing organizations. It is clear that the operational managers as a group have very stable employment histories. Most of them have only ever worked for one employer, or at one facility.

The average level of experience across all interviewees is approximately 25 years. For most interviewees this is their entire working life, indicating a long and stable relationship between employer and employee, and often within the operating team. Ownership of two of the participating sites had changed several times during the working life of the operational managers. The tone of their descriptions was that senior managers come and go, but they, and the facilities, remain. As one interviewee at the nuclear power station explained: 'I think one of my colleagues down the bottom [in the control room] talks about time scales in quarter centuries … that contrasts with a lot of people who seem to do three or four years here and then move on or three or four years there and move on.' This sense of permanence as management comes and goes contributes to a professional sense of independence and autonomy from the mores of each new owner.

In contrast to modern employment practices where people are told to expect to have several different 'careers' over the length of their working life, professionalism has been historically associated with a sense of vocation, a long term, stable commitment to one field of endeavour. This is certainly the occupational path that the operational managers have chosen. In each case, they have reached the top of their profession. Further promotion within their employing organization would entail a significant change to the nature of the role – moving off shift and away from day-to-day operational responsibility to a role with a stronger management focus. This would be a major change in career focus from achievement to advancement as described previously.

The long-term perspective on operations gives a historical, almost archaeological, multi-layered view of the facilities and how they operate. Such a view is not available to people with less experience. A comment by one of the Shift Managers at the chemical plant (Interviewee 2) illustrates this. He was called at home by the on-duty Shift Manager for advice about a plant problem. He said,

> Just by chance I just had the answers in my head and we went with those … I've been here since day 1 … I had the advantage of seeing the plant grow from the ground and there are things that you see that other people don't. They are there for them to see but because of what else is happening around them they don't necessarily see them in the way that you see them.

Other operational managers made similar comments about their technical knowledge. Most of the operational managers have literally grown to professional maturity along with the technical systems with which they work. Several were present when the original facilities were commissioned, giving them a uniquely multi-layered view that cannot be replicated by a newcomer to the system, no matter how much technical ability they bring to their role. Such a depth of knowledge and sense of vocation are professional characteristics that are hugely valuable in ensuring safe operations.

6.5 Public Trust

Another theme evident in the research data is the sense of public trust felt by the operational managers. This takes two different forms due to the varying natures of the primary tasks of the organizations. At the air navigation service provider, there was a strong focus on public service as the operational managers have repeated direct interactions with other members of the aviation community. Operational managers at the chemical plant and the nuclear power station have no direct interaction with customers or members of the public, but they have a strong sense of their responsibility for public safety. This was also shared by the operational managers at the air navigation service provider.

A sense of trust and integrity is another feature of the behaviour of professionals. It is not something mentioned or discussed in the context of organizational safety (although of course it is implied in organizational goals).

6.5.1 Public service

As described in Section 6.2, operational managers at the air navigation service provider are the only ones who deal directly with the general public. Uniquely, the professional identity of the operational managers is therefore focused on public service. There are many small occasions on which staff go out of their way to assist members of the aviation community. Several examples were observed. On one occasion in Sydney, a student pilot become disorientated and lost. The air traffic controller helped the pilot land safely, not only by providing navigation details, but also reassuring her at length over the radio when she started to panic. The operational manager supported the controller by ensuring that other tasks in his area were managed. He also rang the flight school to advise them what was happening to their student and where their aircraft would be landing. A similar event occurred in Melbourne Centre with a foreign private pilot in a small Cessna flying from Alice Springs to Cooper Pedy. Part of this route is outside radar coverage and the air traffic controller who spoke to the pilot as he left Alice Springs was not confident that the pilot really understood Australian flying conditions. Again, the air traffic controller reported his concerns and supervisors ensured that his workload was managed so he could monitor the radio in case of any emergency signals from this aircraft. Many similar almost trivial examples of small acts of assistance collectively occupied a significant part of the operational managers' time.

In both Sydney and Melbourne, operational managers regularly took phone calls on the public telephone line with messages to pass on to pilots of small commercial aircraft. In Melbourne, supervisors spent a significant proportion of their time dealing with signals from emergency beacons. Individual controllers may detect the signal from a beacon that has been activated (or take reports from commercial aircraft crew who have detected it). The individual reports are handed to the operational manager, who must decide on the appropriate action and notify the national search and rescue agency. The overwhelming majority of signals received are false alarms – people inappropriately testing beacons, setting them off by accident, nuisance triggering and faulty equipment – but all signals are treated seriously until the source is identified.

In a similar way, there were many examples of consciously looking out for the best interests of larger customers such as the commercial airlines. As anyone who flies regularly will know, airlines are quick to blame air traffic control for schedule delays, but in fact all personnel do their best to ensure that flights remain safe and on schedule, despite weather or other environmental factors. Many staff volunteered that they were very aware of airline costs (and aware that their managers wanted them to be aware of airline costs). Traffic Manager 1 said that he feels financial pressures acutely, meaning that he feels it is his job to keep airline costs down by such measures as optimizing fuel use and minimizing delays. Feedback from the airlines is mixed. Whilst they did sometimes receive some grumbling, it was generally felt that airline operational staff know that controllers and operational managers 'work hard to fill every available slot'.

Whilst most of us are unaware of it, a network of dedicated people look out for us as we take to the skies. This applies not only to our physical safety, but also to generally ensuring that the complex aviation system runs smoothly and that we all get where we want to go – safely, in an orderly manner and expeditiously.[1] The organization that the managers and controllers work for was oddly absent from the discussions. There was little discussion of corporate goals or requirements. Consistent with this general atmosphere is the story told several times that annual staff satisfaction surveys consistently show a strong loyalty to the profession and a relatively weak loyalty to the organization.

6.5.2 Public protection

The second aspect of holding public trust is the responsibility that operational managers in all organizations feel to protect the public and their fellow workers from danger. Operational managers were keenly aware of their responsibility for safety, as shown in the language they use to describe their thoughts, aims and actions. For many people, metaphors are simply a literary device but the role of metaphor in everyday speech can be seen to be much more fundamental. Lakoff and Johnson (1980) maintain that our cognitive processes are profoundly metaphorical in nature. Metaphors are a part of everyday conversation that impact on the way that we think and act. If we take this perspective on the relationship between language and thought, then the words that we choose to use are 'tools for coping rather than tools for representation'. (Weick 2006) Metaphors used by operational managers provide a clue as to how they see themselves and their role in relation to the organizational system in which they find themselves.

A recurring metaphor used across all organizations studied to represent the plant or physical system for which they are responsible is 'the beast'. One person at each site used this term to describe the technical system with which they work and how they decide whether a given situation is safe. At the nuclear power station, one operational manager said:

> People who say they have an idea of the power of what's in here don't know what they are talking about. The power is just mind boggling. It's beyond my comprehension. As we say, our job is to keep the beast in the box.

In describing plant safety performance since a serious incident some years earlier, an operational manager at the chemical plant said, 'that [equipment item] has

---

1   Air traffic control is defined by the International Civil Aviation Organization (ICAO) as 'a service operated by an appropriate authority to promote the safe, orderly and expeditious flow of air traffic'. The phrase 'safe, orderly and expeditious' is used commonly in the ATC field.

failed since and the beast was able to be [shut down safely]' and in describing his role, one operational manager working in air traffic control said, 'you need to learn about the beast and understand the interactions and impact on operations'.

In the case of the nuclear power station, the beast referred to is the nuclear reactor itself. Similarly, for the chemical plant, the beast is also a reactor, the most hazardous part of the plant. For the air navigation service provider, the beast is the physical system that allows the controllers to do their work. Looking at the quotes in context, the qualities of the beast are common to all three interviewees. In each case, the beast is:

- Complex,
- Unknown and to some extent unknowable, unpredictable,
- Contained, but difficult to control,
- Powerful, and
- Dangerous and even somewhat malevolent.

These qualities of the technological systems were also reflected in other stories that did not use this specific metaphor. As one chemical plant operations person described to me recently, the plant where he works 'needs to be coaxed, looked after, pampered and even spoilt – or otherwise it will throw a fit, get angry or even bite you'. The implication is that, provided the beast stays in the box, things are safe – 'That's how we operate [using the conservative decision-making practices detailed in Chapter 3]. Because we understand the complexities of the plant, the inherent potential danger of the plant if it weren't to be run safely'.

So how do operational managers see themselves in relation to the beast? This metaphor perhaps conjures up images of a heroic character, a brave, isolated figure taking personal risks to battle against, and ultimately defeat, the beast to the accolades of lesser mortals. This heroic view of a battle against the beast is sometimes used in the context of safety, as is typified by the headline from the *Adelaide Advertiser* newspaper in early autumn 2008 which said 'Strike Force – how the CFS conquered the beast'. (Robertson et al. 2008) The headline referred to how members of the volunteer fire fighting service (the Country Fire Service or CFS) had conquered the rampaging beast, in this case a series of bushfires, at great risk to themselves and hence averted disaster.

Few, if any, of the 23 stories recounted in earlier chapters have this sense of the heroic. The operational managers operate as part of a team which operates conservatively, rather than as extraordinary or superhuman individuals, and there are few metaphors of battle or fighting used in the stories told by interviewees. One rare exception was an operational manager at the nuclear power station who said in relation to a serious water leak: 'OK enough is enough. We can do so much on this. We can either attack it on a repair front and we're also now trying to attack it on a contingency front.' Even in this case, the manager described his actions as operating as a member of a group (which is not typical heroic behaviour). His considered language is in marked contrast to the language used in the article discussed above where the fire fighters' 'ferocious attack' and their ability to 'strike quickly' had 'averted a series of potential disasters'.

Praise for heroic behaviour is common in many organizations, not just in relation to safety, but also for solving urgent problems generally. Heroic tales of organizational successes abound – where an individual or small close-knit team, acting against all odds, manage to pull off a remarkable feat that saves the project, if not the entire organization, from some imminent disaster. In many organizations, this type of behaviour (often informally described as 'fire fighting') would receive high praise from management, despite the fact that better planning may have prevented the situation from arising in the first place. It appeals emotionally to our sense of the dramatic and also to our rational problem solving abilities.

In a similar vein, Woods (2006) describes what he believes to be the basic story told within organizations attempting to foster a good safety culture:

> Someone noticed that there might be a problem developing, but the evidence is subtle or ambiguous. This person has the courage to speak up and stop the production process underway. After the aircraft gets back on the ground or the system is dismantled or after the hint is chased down with additional data, then all discover the courageous voice was correct. There was a problem that would otherwise have been missed and to have continued would have resulted in failure, losses, and injuries. The story closes with an image of accolades for the courageous voice. (Woods 2006: 31)

Woods points out that this story has the wrong ending and that an organization that was truly resilient would champion courageous actions to interrupt production where further analysis showed that there in fact was no problem and that, in an immediate sense, the shutdown was not necessary. This is certainly a valid point and the way in which organizations and operational managers judge the quality of safety decisions is addressed in Chapter 8. But another serious problem with the story told as a safety archetype is the heroic nature of the actions of the individual.

This behaviour is the antithesis of that advocated by the HRO theorists (see Chapter 2). The HRO approach favours organizational learning so that problems are identified and solved before they need dramatic, and potentially heroic, interventions. Taking responsibility for seeking out potential problems and acting early is an orientation required of all staff. Schulman (1996) found in his research that the attitude of personnel at the Diablo Canyon nuclear power station was not heroic but he found that staff at an oil refinery did favour and admire heroic behaviour (telling stories of individuals opting to go it alone in problem solving). Chapter 8 includes much more discussion about the stories told by operational managers, but these did not have the form of heroic tales. No interviewee thought swift, unilateral action was an effective operating strategy.

Operational managers took a different view of their relationship to 'the beast'. Much of the language used by operational managers in talking about their general attitude to safety relates to their need to guard or protect others. Some examples from people at the nuclear power station:

> I know what I need to be doing to ensure my own safety and the safety of my team members as well and ensuring that they can continue to work safely.

Also

> You can always put it back [that is, start up again]. That's the point. If you hurt somebody or you make a big mess of the machine or you put the press into overdrive about how dangerous these reactors are and the public don't like it, then that doesn't help anybody, does it?

And from people at the chemical plant:

> I think the safety of my crew was the thing that was uppermost in my mind and I wanted to make sure that that was the thing we took care of before anything else, so I certainly wasn't about to put anyone in an unsafe situation.

And

> I would never ask [someone to do] anything that was unsafe in any way.

It is perhaps worth noting that at the nuclear power station and the chemical plant, the people impacted by any potential incident were the workers themselves (and in the worst case, people around the site). The situation at the air navigation service provider was somewhat different in that the workers' safety was not directly impacted by incidents they were aiming to prevent. Nevertheless, there was no noticeable difference in their attitude in this respect. None of the supervisors expressed any fears for their own physical safety. Their fears were around their responsibility for the safety of others and informing loved ones of an injury or death.

Some people were particularly aware of the potential impact of any incident on their workmates and the job that they themselves might have in justifying their actions to relatives if something were to go wrong.

> I think it's like the ads on the telly recently.[2] You don't send people out to work and not come home the way they went. I see the role that I've got as pretty critical to that happening for most of the people who are under me. I mean, signing on permits and the operators working safely, doing what they are supposed to do,

---

2   This is a reference to WorkSafe Victoria's television advertising campaign which shows a young boy waiting anxiously for his father to come home from work.

following instructions as they should be. You try your hardest and hopefully you don't send anyone home injured.

The underlying common conceptualization of the role of the operational managers was one of guardian or protector.

6.6 Summary and Concluding Comments

This chapter has aimed to demonstrate how the overall attitude of operational managers to safety and production issues is impacted by their view of themselves and their peers as both organizational members (or employees) and as professionals. Neither identity is universally good or bad for safety decision-making. Each has a range of characteristics that, given the right circumstances, can contribute to good overall safety performance.

The existence of a conflict between professionalism and managerialism has a long history in academic publications on the theory of work (see Causer (1996) for a summary). As early as the 1960s authors highlighted two potential sources of conflict: firstly, that professionals may see themselves as more closely aligned to public service than the organization for which they work; and secondly, that managers may wish to constrain the independence and judgement of the professionals under their direction by imposition of bureaucratic rules. These features have both been seen in the organizations studied, and certainly these points of conflict have highlighted the fact that operational managers have a strong professional identification in addition to their organizational affiliations.

On the other hand, simply to characterize their dual occupational identities as an issue of conflict is to oversimplify the situation. On most occasions, operational managers work effectively as both employees and professionals. There has been a strong focus in safety performance improvement in recent years on the role of organizational leaders in encouraging safe behaviour from all staff. This has emphasized the need for leaders to constantly reinforce safety as the number one goal of the organization, with production always in second place. This approach to organizational safety makes an appeal to each individual in his or her role as a member of the organization. Section 6.2 shows how organizations define the role of operational managers and delegate responsibilities so as to encourage the right balance between safety and production.

The research data suggests that operational managers have a professional affiliation that also has the potential to influence their attitude towards safety and the sense they make of any specific situation. Section 6.3 shows that operational managers have a strong professional pride in production that can sometimes influence their decisions to continue to operate, irrespective of stated organizational priorities. Section 6.4 describes the vast technical experience and hence multi-layered view of operations and sense of vocation that operational managers bring to safety decision-making. They see problems and solutions that may not be apparent to those with a flatter view of the system. Section 6.5 describes the sense of responsibility that operational managers feel for the safety of the general public and their colleagues.

Compared with discussion of organizational priorities and compliance with rules, these professional issues receive little organizational attention. Organizations that ignore these factors may have missed an important opportunity for influencing attitudes to safety. The implications of these overall attitudes for safety decision-making are discussed in Chapter 8.

Chapter 7
Rules and Compliance

The human ability to adapt and innovate is both a strength and a weakness when it comes to operating complex systems. Safe operation requires a balance between the need to behave reliably in prescribed ways when things are operating in a way that the system designers foresaw, and the need to use judgement and experience to find a creative solution in a situation that no-one has previously seen or imagined. Rules and procedures play a key role in finding the right balance. This chapter explores the way in which rules both constrain and support operational managers when making safety-related decisions.

There are many circumstances in which the appropriate action to ensure continuing safe operation in a given situation is something that can be completely specified in advance in the form of a rule. This is essentially the view of the quality management theorists whose influential approach now forms the basis of regulations, industry standards and company procedures which all focus strongly on compliance with rules as a means of managing and controlling operations to ensure safe performance. The chapter begins by looking at the background to the quality systems approach and its weaknesses as well as its strengths for management of operational safety in complex systems.

As Reason (2008) describes, people can be seen as either hazards (an unreliable component in the system that needs to be controlled) or heroes (capable of great adaptability and recovery, and great enhancers of system performance). If people are hazards, then rules need to be about controlling them. Alternatively, rules can be designed to inform and support decision-makers. Section 7.1 further develops these alternative views of the role of rules in reviewing the purpose of rules for safety decision-making. The remainder of this chapter focuses on two different types of rules that are important for operational decision-making. Section 7.2 discusses use of specified engineering limits in system operations. Section 7.3 describes decision-making procedures within the defined operating envelope.

7.1 Quality Management and Safety

In the twenty-first century, safety management systems are ubiquitous in high hazard industries. At their best, they are relevant, accurate, available, easy to use, tightly controlled and yet easy to update. Many of us have worked with systems that do not meet aspects of this description and are patchy, verbose, difficult to use, or just plain wrong in significant ways and, as a result, collect dust in corners of offices and control rooms.

Readers will have realized that this is not a guide to writing your safety management system. Many excellent industry standards and guidelines are available and, for many jurisdictions and industries, compliance is mandated in regulation. Such standards and guidelines are almost always based on the ideas of quality management. Since rules and procedures are now such a well-accepted way of ensuring good safety outcomes, it is relevant to revisit the origins of this approach and to review what research tells us about some of the underlying assumptions.

Modern management techniques have their origins in the activities of nineteenth century railroad companies. Prior to the development of large scale railroad networks, most commercial organizations were quite small, with less than 50 staff. Management activities were undertaken by the owners of the business and many organizations relied on outworkers doing piecework at home and working semi-independently. When rail networks were developed for the first time significant activities taking place over a wide geographical range needed to be coordinated. Following a significant accident in 1841 that resulted in two deaths and 17 injuries, the Massachusetts legislature launched an investigation (Chandler 1977). The resultant report made a series of recommendations regarding the need to better define responsibilities, lines of authority and communications between various parts of the organization. This incident and others provided the incentive for development of professional salaried managers, a new group distinct from owners or workers. Most managers had originally trained as civil engineers, because the management staff developed from the technical specialists involved in building the new systems.

Improvements in transportation and communication revolutionized manufacturing industry, making it not only feasible, but also economic, for organizations to specialize in particular components and for finished goods to be made from parts manufactured in widely differing locations. In the late nineteenth century, the members of the newly formed American Society of Mechanical Engineers turned their attention to management of manufacturing organizations. In 1895, their combined experience was first reported on and expanded by Frederick W. Taylor to describe what was to become known as scientific management. Taylor soon became known as the foremost expert in factory management, employed by firms such as Du Pont and General Electric to advise on organizational structures and cost control (Chandler 1977). Morgan (1997: 23, italics in the original) summarizes Taylor's principles as follows:

1. *Shift all responsibility for the organization of work from the worker to the managers.* Managers should do all the thinking related to the planning and design of work, leaving the workers with the task of implementation.

2. *Use scientific methods* to determine the most efficient way of doing work. Design the worker's task accordingly, specifying the *precise* way in which the work is to be done.

3. *Select* the best person to perform the job thus designed.

4. *Train* the worker to do the work efficiently.

5. *Monitor* worker performance to ensure that appropriate work procedures are followed and that appropriate results are achieved.

Taylor's work on scientific management has been hugely influential in management theory throughout the twentieth century. His work is based on the assumption that workers, with appropriate training, are interchangeable components in the system. Much modern management theory is also based on the metaphor of organization that sees the workers as cogs in a large machine. Quality management originated in the 1970s and has its origins in the same tradition. This approach was originally popularized by influential American theoreticians Deming and Juran working within the Japanese automotive industry before being adopted in the US. It was originally the domain of manufacturing organizations with a business need to produce large numbers of identical, but relatively simple, items. The basic idea behind this management philosophy is that the key to improving the quality of the output of an organization (and hence the key to customer satisfaction and ultimately the success of the organization) is definition of, and standardization of, work processes. In recent decades, this management philosophy has spread from manufacturing throughout engineering-based industries and has formed the basis of other popular management tools, such as business process re-engineering (often known as BPR) (Beamish 2002).

Organizational control is exerted via the well-known Deming cycle or feedback loop consisting of four steps – Plan (establish goals and objectives), Do (implement processes), Check (measure output against goals), Act (act to improve performance) – which are seen as sequential, cyclic stages that operate at every level of the quality system, from individual processes to the system as a whole. There are obvious similarities between the PDCA cycle and Taylor's management principles.

Whilst Deming and Juran (and other more recent quality theorists) have also emphasized organizational learning (Deming 2000) and the importance of culture in adopting new business processes (Defoe and Juran 2010), quality management maintains its focus on business process mapping and control as the key to performance improvement. These ideas have been responsible for major improvements in organizational production efficiencies – faster production of higher quality products with less wastage – but quality management does not necessarily mean safety management. Prior to the loss of the *Columbia* space shuttle in February 2003, NASA management had adopted an organizational philosophy based on quality principles that they called 'faster, better, cheaper'. Ocasio (2005) points out that the concept of 'better' as promoted within NASA did not include seeking improvements to safety. The goal of organizational safety was external to this management philosophy and was spoken of as 'a constraint to be observed rather than a goal to be pursued'. (Ocasio 2005: 108) This kind of thinking was a contributor to the circumstances that led to the loss of the space shuttle.

It is common to attempt to integrate management of safety and quality, or at least to apply the same management approach. Safety management system standards often advocate the same cyclic approach to performance improvement as the quality cycle described above (see for example (ILO 2001) and (Standards Australia 2001) which are both based on the quality management system standard ISO 9001 (Standards Australia 2008)), and no doubt implementation of management system principles has helped many organizations focus on and improve their overall safety performance. But preparation and implementation of a written system for managing safety is not the universal panacea that some people believe. This model (consistent with the overall quality system approach) assumes that:

- Safety can be achieved by goals specified in advance, with no specific consideration given to the fact that goals may be conflicting,
- The link between goals and the processes required to achieve them is knowable and largely static, and
- Learning via a feedback loop and making corrections to the system based on less-than-ideal performance, is an adequate strategy for improvement.

Even in complex systems, much activity is routine and a system of work based on these assumptions is an effective way to ensure basic standards are met. But these assumptions are all somewhat problematic if they are seen as a sufficient strategy for managing all aspects of safety. Strategies are required that can deal not only with conflicting goals but also with unanticipated events. Trial and error learning is not a sufficient learning and performance improvement strategy when the consequences of error are potentially so great.

So where does this leave us? In the quality approach (at least as it is often implemented), procedures are a form of management control, imposed on workers by those with a greater knowledge of what makes for efficient (or safe) operations. Whilst a level of control is necessary and desirable (to ensure, for example, that design constraints are carried forward into operation of the system) complex systems often perform in ways that require discretion and judgement on the part of the operating professionals. This leads to a rethink about the purpose of written procedures, especially as they apply to operational managers. The metaphor of control is not the only way that rules can be conceptualized. An alternative view is that rules should be seen as support and assistance for decision-makers. The proposition that senior managers should control the activities of the operational managers conflicts with the concepts of professionalism described in the previous chapter. '[F]reedom of judgement or discretion in performing work is intrinsic to professionalism, which directly contradicts the managerial notion that efficiency is gained by minimizing discretion'. (Friedson 2001: 3) This is also consistent with concepts of high reliability where deference to expertise has been identified as an important factor (see Section 2.1).

In this alternative view, rules are best seen as a communications tool – a way of ensuring that operational managers are aware of fixed constraints that are outside their field of expertise and also of broader organizational expectations. This encourages those who write rules to see them as a means of supporting and assisting operational managers, rather than a method of controlling them. Whilst the metaphor of rules as a form of control has some validity, it encourages those who write rules to believe that they have a complete understanding of the system that they are controlling remotely. This is simply not the case, as an air traffic controller explained in the popular press (Hollingworth 2007):

> Some days everything is on the rails; it's predictable and the weather is benign. Other days, everything goes wrong. A couple of nights ago, we had big southerlies come through and that can cause bedlam. We had six or seven aircraft that couldn't land because of turbulence, which meant our system was overloaded for half an hour. You don't have any rules on how to sort it out; you have to rely on your technique and your experience. It's more of an art than a science.

In this article regarding how people in stressful jobs relax, he contrasted his working environment with a different environment that is knowable and where following rules always gives a predictable outcome 'I also like to cook … in the kitchen I can follow a recipe and the temperature is set. It's all controlled and it all goes to plan'. In this case, he is literally contrasting air traffic control procedures with a cook book.

Compliance with procedures that accurately translate design limits and past lessons learned into plant operating constraints is important for safety. But more is needed from operating teams than simple compliance (or we would surely automate those procedures and eliminate the need for those potentially unreliable people). Organizations that (consciously or otherwise) see management systems in Taylorist terms, that is, primarily as a method of management control, are ignoring important safety performance improvement mechanisms.

This message is consistent with safety culture research, too. Westrum initially proposed a three stage progression in improving safety culture from pathological to bureaucratic and finally to generative (Reason 1997: 38). Patrick Hudson's development of this work into a five-stage model describing the evolution of safety culture is well known. In this scheme, a culture focussed primarily on a command and control, top down management style, with safety performance in the field believed to be driven by compliance with procedures, is the third of five possible stages of development. A more effective safety culture is described as one where 'an internalised model of good practice' (Hudson 1999: 8.6) becomes the driver for action.

7.2 Implementing Effective Operating Limits

Having considered what the purpose of rules might be and why organizations think rules are so important, this section and the next look at two important types of rules that relate to system safety. The first is the concept of operating limits, system boundaries, critical operating parameters or the safe operating envelope (there are many terms for this set of parameters). Safety decision-making at the nuclear power station was strongly influenced by compliance with such an operating envelope defined in what were called Station Operating Instructions (SOIs). The limits fixed in the SOIs typically take either of two forms:

- The maximum or minimum allowable value of a plant operating parameter such as pressure, temperature or concentration (for example maximum reactor core temperature or maximum oxygen concentration in primary coolant), or
- The minimum operational level of redundancy of safety-critical equipment for example must have at least three of five pumps operating or available.

Due to the complexity of the operating plant, there is a series of SOIs covering different parts of the facility (several dozen in total). The SOIs do not prescribe or even suggest actions that may prevent the operating limits from being breached. They simply define the limits of safe operation of the facility and give the operating crew, under the direction of the operational manager, the freedom to operate within that envelope. Breaches, or potential breaches, of the limits set in the SOIs are not tolerated and a reactor will be shut down if there is seen to be no other way to avoid approaching one of the specified limits (see Story 2 and Story 3). SOIs are also used in maintenance work planning, as described in Story 9.

SOIs are formulated by engineering considerations outside the operational managers' direct experience and knowledge. For example, a particular operation may require two pumps running out of the five pumps installed for this service, in order to meet numerical reliability targets for the system). The safety rule might further specify one extra pump on standby (in case one of the on-line pumps fails) in order to meet the required reliability of the system. The overall requirement might be therefore a minimum of three pumps available at all times. The operational manager might then use this rule to decide whether a pump can be taken out of service for preventative maintenance on a given day. The requirement fixing the number of pumps will have been determined in turn by the overall reliability required of the system. This example may seem straightforward but, in complex cases, such considerations are based on complex reliability calculations by specialist engineers linked to quantitative risk calculations of the type described in Chapter 2. An operational manager does not need to understand these details in order to do his job.

Similarly, many of the SOIs define process conditions such as maximum operating pressure, minimum operating temperature or maximum concentration of a contaminant. These are almost always fixed by design considerations, such as the integrity of the process equipment or the response of the process itself to severely abnormal conditions. These issues are the domain of specialist engineers, not operations staff. Fixing such limits is the primary way that these design considerations are transferred from one group of specialists (those who designed the facility) to another (those who operate it). In this way, SOIs provide a clear boundary beyond which continuing operation is unsafe.

So in this way, Station Operating Instructions are Type 1 or goal-based rules (Hale and Swuste 1998). This type of rule gives the highest degree of freedom to the decision-maker. It specifies only the general outcome required and leaves the details of how the goal is to be achieved to those doing the work.

This is one of three distinct types of safety-related rules. The other two are:

- Rules that define the process to be followed in order to decide on a course of action (Type 2 rules),
- Rules that define a specific concrete action or system state (Type 3 rules).

Type 2 rules (process-based rules) describe the sequence of steps that the decision-maker is required to complete before coming to a decision about the course of action required. In this case, the detailed outcome is not specified (although a general goal is usually inherent in the context of the prescribed process). Many work management systems are process-based rules (for example, a permit to work system or a requirement to conduct a job safety analysis). The basic assumption behind such systems is that, by involving and informing specific people, and prompting them to consider a range of specific questions in relation to the planned activity, work will be carried out in a safe and timely manner.

Type 3 rules specify tightly the behaviour required of an individual. They involve much less interpretation than the other types of rules. Examples include hard and fast requirements to wear specific protective clothing whilst undertaking certain activities or requirements for staff to be licensed in order to carry out certain tasks. Detailed operating procedures are also Type 3 rules. These three types of rules represent an increasing limitation to freedom of choice or a tighter degree of control and specificity.

The difference between these differing types of rules in practice can be illustrated by the approach to specifying operating limits in two different industry standards. The nuclear industry standard (International Atomic Energy Agency 2000) advises organizations to take a multi-faceted approach involving all three types of rules. They are:

- Detailed rules to cover normal operations and response to process alarms (Hale's Type 3 rules),
- Defined operating limits that come into play if the actions described in the detailed rules do not bring the plant back into the desired state (Type 1, or goal-based rules), and

- Rules defining a process to follow to assist decision-makers in abnormal situations within the operating limits (Type 2 or process-based rules).

A suite of procedures designed along these lines would seem to provide an excellent balance between uniformity and innovation. On the other hand, one influential chemical industry standard (Center for Chemical Process Safety 2007) adopts a system of managing operating boundaries based firmly on Type 3 rules, where the required actions are totally specified in advance in all cases. Clearly, this assumes all possible circumstances can be anticipated and takes responsibility for decision-making away from operating personnel in a way that is based on scientific management principles described earlier. This may be appropriate in managing some processes but it brings with it the problems inherent in that approach described in Section 7.1.

The air navigation service provider also uses a set of operating limits to link required system reliability and operations/maintenance activities, but it operates in a rather different way to the system at the nuclear power station. When equipment items fail, the urgency with which they must be repaired is specified by the nominated maximum time to repair – the System Restoration Time or SRT. SRTs have been developed based on the cost to repair, balanced against the increased risk to aircraft and passengers of the item being out of service.[1] There is a very high level of redundancy in the equipment systems used by air traffic controllers to monitor and talk to aircraft, so failure of a single item does not usually result in any immediate reduction of the air traffic control service available. The potential safety issue is that, until the item is repaired, there may be no remaining redundancy in the system, so that a further failure may lead in some cases to abrupt and unplanned service interruptions. The time taken to repair failed items therefore can have a direct impact on safety margins.

The time taken to repair also has a significant impact on cost for this highly distributed system. For some items in remote locations (such as radar beacons scattered across the country), a short SRT might mean a special trip (even chartering a plane), whereas a long SRT might mean fixing a fault on the next scheduled visit at a much reduced cost. SRTs have therefore been determined to balance the cost of the repair against the reduced safety margin resulting from reduced redundancy in the system.

On the other hand, it is acknowledged within the system for managing SRTs that not all situations for repair of a specific item are appropriately covered by the general rule. There is a system in place where exception reports can be issued by

---

1   This is an operational view of the concept of SRT. From a broader system design perspective, further cost trade-offs must be made between system design options, the risk associated with system failure and the time to repair i.e. a system may be made more reliable, instead of accepting failures and repairing them in a particular time frame. In practical terms, for day-to-day operations the design is fixed and the trade-off becomes repair cost versus safety.

operational managers, either requesting a shorter restoration time or authorizing a longer time. Story 20 is one such case. This means that, whilst SRTs give an indication of the required performance of the maintenance system, they are not treated as hard and fast limits in the way that SOIs are treated at the nuclear power station. Given that the SRTs are typical, rather than maximum, acceptable repair times, allowing (or even encouraging) the use of exception reports would seem to be a good thing for safety, so that repairs can be done more quickly when necessary.

The potential problem is that this system creates a way to disguise problems. Performance indicators for the maintenance group include minimization of both overtime costs and the number of occasions on which restoration times are not met. Operational managers are encouraged to issue exception reports if savings can be made by extending restoration times in particular cases, provided they see no compromise to safety. These factors combine to put the operational managers under pressure to issue exception reports to extend the allowable restoration times if maintenance overtime would be required to meet them. These decisions required from the operational managers represent a direct trade-off between a small reduction in safety margins and the cost of overtime. This is appropriate and to be expected but, in this case, the need for such a choice and the decision made are hidden and not available for review since records are based on the number of occasions on which restoration times have not been met. The original driver for development of SRTs was to define an internal agreement between operations and maintenance as part of an overall corporate move to contestability and potentially tendering maintenance services. SRTs have a commercial flavour that is not present in discussion of SOIs at the nuclear power station. SRTs are seen as an operational guideline, whereas SOIs are fixed safety rules.

In summary, setting operating limits based on engineering considerations to guide operations people in application of their own experience and judgement is important for safe operation. There is strong support in regulation and in industry standards for this approach. Such a system of rules is in place at the nuclear power station and it is both well understood and respected by operational managers. Explicit consideration should be given as to whether procedures should include instructions on how such limits should be maintained (and actions to be taken in the event that they are threatened) and to what extent these decision should be left to the operations personnel.

The experience at the air navigation service provider suggests that care should be taken in describing how such limits should be used. Confusion about whether the limits are desirable, typical values or fixed limits not to be exceeded must be avoided.

7.3 Setting a Line in the Sand

So far, we have seen that, in complex systems, the best safety outcomes are achieved when it is acknowledged that not all circumstances can be anticipated and that on occasions the experience and judgement of operational managers will come into play. Under these circumstances, written procedures can still be a useful support to decision-making processes, as is seen by the following example from the nuclear power station. This documented system for conservative decision-making is an example of a process rule – a rule specifying a process to be followed, but leaving the outcome to the discretion of the decision-maker.

At the nuclear power station, conservative decision-making is stated in written procedures to be the required attitude or general orientation to decision-making. In practice, operational managers had independently developed a process rule that they uniformly applied. Interviewees described that, within the bounds set by the Station Operating Instructions, setting decision-specific limits and sticking to them is a core component in the station's philosophy of conservative decision-making. Under this approach, as an abnormal operating situation develops, the operational manager considers the available information and fixes a situation-specific limit beyond which the facility will be moved to a safe state (usually shut down). This is similar to an operational limit imposed by an SOI, but it is something particular to the operating situation at hand. The limit is often time, as in Story 5, where the operational manager fixed a time limit before initiating a reactor shutdown when the operating crew was trying to control a significant water leak (not directly related to reactor cooling). Sometimes other process parameters are used, as in Story 8, which described how problems on startup were managed. Within this self-imposed limit, the operating crew continue to monitor the situation and attempt to solve the problem. If the limit is approached, then activity moves from troubleshooting to shutdown. The operational managers at the nuclear power station are collectively very proud of their approach and have had some years of practice at setting, articulating and abiding by the relevant 'line in the sand' for unusual operating conditions.

In contrast to this, at the chemical plant, operational managers (along with other technical staff, including engineers) had received training in a system of decision-making that follows the five steps known as classical decision-making – define the problem, identify possible solutions, evaluate the solutions, implement the chosen solution, check if the problem is solved. This approach had been formalized in a procedure but the operational managers completely ignored it to the point where, in approximately 15 hours of interviews, none of them mentioned its existence. When asked why this might be the case, the responsible manager said: 'well it's really just what you should do out there. I think a lot of them do [make decisions in this way], but they don't have those words going around in their heads so they don't always recognize that is actually what they do'.

Whilst senior management may be of the view that the system simply formalizes what the operational managers already do, the managers themselves do not agree. In fact, several of the stories told by operational managers at the chemical plant followed a pattern that had similarities with the 'line in the sand' approach adopted by their counterparts at the nuclear power station. In one case, the manager was involved in temporary repairs to a leaking cooling water system. Since the plant was still operating, he had set the control room operator the task of monitoring a plant parameter with instructions to shut equipment down if a specific limit was reached (see Story 14). The difference at the chemical plant was that there were also several stories told where the limit set in the first instance was then ignored as repairs were delayed for a range of practical reasons (see Story 15 and Story 16). In some of these cases, the situation deteriorated and an emergency trip (rather than a controlled shutdown) resulted. The stories show our human tendency to say, 'well it's been OK so far' just like the frog in the pot of hot water mentioned by one of the operational managers at the air navigation service provider. In some cases, we revise our original view that the activity or operation was undesirable. This leads to acceptance of continuing operation with decreased safety margins.

In her study of the *Challenger* disaster, Vaughan (1996) found that, over a period of years, NASA technical staff came to accept observed damage to solid rocket booster seals as normal, even though it was initially seen as a problem. Eventually the seals were so damaged on one launch that they failed and the shuttle was lost. She calls this shift in what is normal or accepted practice 'the normalization of deviance'. This work shows that such normalization can occur very quickly in cases where the self-imposed limit is not strongly articulated and/ or recorded.

Similarly, the operational managers at the air navigation service provider were observed creating situation-specific rules to limit operations. Story 21 describes a case where a revised weather forecast was received predicting significant deterioration during the afternoon. The operational manager immediately reworked the schedule of flight arrivals and departures for the airport, using a reduced cap of the maximum number of movements per hour. He chose the cap based on his experience of past similar situations – what it would be possible for the controllers to manage in the conditions of weather and other environmental factors likely to exist later in the day. The new arrangements for the afternoon were distributed to the airlines and the air traffic controllers, so there was no question that the new situation-specific limit could be ignored. This figure chosen by the operational manager based on his experience simply became the new planning basis for the day.

It seems that the 'line in the sand' approach has been accepted as a process rule because it supports the cognitive processes that the operational managers naturally use as experienced decision-makers. This approach does not dictate to the operational manager how he should come to a conclusion about the safety or otherwise of the system. Rather, it specifies a way of making him stick to his judgement once he has drawn initial conclusions (unless the situation changes). The alternative model supposedly in place at the chemical plant attempts to dictate the sequence of thought processes that should be used in reaching a conclusion. It does not follow the way the operational managers naturally think about these types of problems, and so they have ignored it.

7.4 Summary and Concluding Comments

Organizations control the actions of their employees (including operational managers) by using rules in an attempt to minimize the hazard posed by those employees. This way of conceptualizing the role of rules is unhelpful for experienced professional people. Replacing staff by automatons that simply follow the rules would not improve safety performance in complex systems such as air traffic control or process plant operations. The human ability to learn, adapt and innovate in unforeseen circumstances saves the day on many occasions. A better approach is to see rules as a way to support professional staff at the limits of their professional competence for example to transfer design information into an operational environment.

There appears to be some misconception within the organizations as to the degree of freedom experienced by operational managers. Senior managers often revealed the view that their operational personnel have relatively little freedom and that their job was largely application of concrete rules. This view is shared and perhaps fostered to some extent by industry standards, regulations and quality systems that emphasize the role of rules and largely ignore the cases in which the rules as written cannot and should not apply. The reality is different. 'Beneath a public image of rule-following behaviour, and the associated belief that accidents are due to deviation from those clear rules, experts are operating with far greater levels of ambiguity, needing to make uncertain judgements in less than clearly structured situations.' (Wynne 1988)

Readers who work in industry will no doubt be familiar with recommendations from incident investigations that focus on updates to procedures. One example is BP's investigation into the causes of the Deepwater Horizon incident which included a series of changes to drilling procedures as the first seven of its recommendations (BP 2010). Of course it is important to ensure that procedures reflect lessons learned from past failures but the fact that procedures are so typically updated following an incident surely also emphasizes that procedures will never cover all possible eventualities.

Of particular relevance to senior operations personnel are system operating limits. Different industries and organizations have differing views as to how such limits are best applied and the case studies have emphasized the potential for confusion and lack of clarity around requirements if limits sometimes take the form of firm values requiring strict compliance and on other occasions are best seen as guidance to preferred values. Within the operating limits, operational managers were observed making their own situation-specific rules. In generating and applying such 'line in the sand' rules, the operational managers rely on their professional experience and are acting from their professional, rather than organizational, identity. This is the subject of the next chapter.

Chapter 8
Professionals at Work

As described in Chapter 6, operational managers have a dual occupational identity – both as employees and as professionals. As employees, they generally follow the requirements of their employer through use of rules, as described in Chapter 7. This chapter turns to the professional side of their occupational identity and explores how this impacts on decision-making.

We have seen that operational managers hold ultimate authority for minute-by-minute operations. This authority is acknowledged in position descriptions, but seems out of alignment with organizational lines of authority (where one would expect the CEO or similar to hold final operating authority), because it is based on professional, rather than bureaucratic, criteria. The operational managers in this study each have vast experience in their chosen professions and exhibit stable employment histories. They have a deep and multifaceted understanding of the complex systems for which they are responsible. Ironically, this high degree of familiarity with the technology also leads them to characterize it as 'the beast' which can still cause unpleasant surprises and must be contained in order to protect the public and their colleagues.

On the other hand, professional considerations are not always conservative with regard to safety and can potentially delay a decision to shut down in the face of a developing problem. Whilst professional goals and organizational goals regarding safety are often coincident, operational managers take pride in their profession – generating electricity, making plastic, moving aircraft – and sometimes feel a self-imposed pressure to produce. Interrupting operations can be seen as professional failure – letting the system get the better of you.

Professionalism is associated with qualities such as loyalty within the profession, a sense of vocation, identification with the goals and values of the profession, integrity and public trust. This chapter builds on those professional characteristics and considers how they impact decision-making.

As described in Section 8.1, experience plays an important role in the ability of the operational managers to see a developing problem amongst the large amount of data available to them at any given time. This is not only about technical knowledge, but also about the ability to distinguish between safe and unsafe situations based on previous experience of both types of situations. This need for contrasting experiences is reflected in the themes of the stories told and emphasizes the difficulty that less experienced staff have in developing a deep understanding of the dangers of the system, without having personally experienced those dangers in some way. The literature also warns that over-exposure to danger can inhibit the ability to work effectively. Good decision-making requires the operational managers to have confidence in their ability to control the situation.

We have seen that stories play a role in allowing operational managers to identify problems. Use of personal stories continues when managers come to choose a course of action. Section 8.2 describes three types of story-based tests that operational managers said they use to determine whether a course of action they are considering is a good one. The tests are based on protecting themselves from the potential for strong negative emotions such as shame, guilt and grief.

With experience playing such a key role in professional decision-making, processes for learning are important. Again, stories are a key factor, and story-based learning is described in Section 8.3.

8.1 Anticipation and Judgement – Seeing the Potential for Trouble

In any specific situation, the first step in making a judgement about the need to interrupt operations for a safety-related reason is to recognize that some kind of problem or potential problem exists. The organizational view of problem identification is generally one of compliance (are any parameters close to defined limits?), but operational managers try to anticipate developing issues and steer the ship away from problem areas. They also may decide that a situation is so unsafe that operations need to be curtailed, even though no defined limit is in danger of being breached. To act in this way requires the ability to see something anomalous in the current situation (a 'cue') and to foresee how the current situation may plausibly develop in ways that are not desirable.

Identifying cues is partly a matter of technical skill. This is one way in which operational managers apply the layered knowledge of facilities that they have gained from literally being present when the facilities and systems were built, and as they have been modified and developed over time. But technical knowledge is not the only thing necessary to be able to notice important cues about the state of the system from the mass of information available at any given moment in a busy plant, control room or operations room. The data suggests there is another important factor in being able to recognize cues. Having experience of unsafe situations seems to provide an important contrast in deciding what is safe.

8.1.1 Safe in contrast to unsafe

> Our motto on the site is 'safe quality tonnes', so if we can't do it safely we don't do it.

> Safety comes before production always. Always.

> Safe, orderly and expeditious movement of aircraft. Safe comes first.

These quotations from three operational managers are typical. All operational managers are adamant that safety is their first priority. Despite this, they were unable to articulate in analytical terms how this general goal translates to a judgement in any specific case. All interviewees were very clear that they could tell whether a given situation was safe or unsafe, but they were unable to describe how they made that judgement or articulate what are the differentiating features. This appears to be because the judgement is based much more on how a given situation makes them feel, rather than on their analysis of it. In attempting to describe how they know that their actions were safe in a given case, several interviewees chose to contrast the current state of affairs with some point in the past that was clearly unsafe in their view. These were not technical comparisons of physically similar situations, but rather comparisons of the way the different situations made them feel.

These stories have much in common with Weick's rich comparisons. He suggests (2007: 17) taking a postcard reproduction of a work of art to compare with the original piece when next you visit a gallery. His contention is that you will notice and appreciate the qualities of the original, partly because of the limitations of the reproduction, in other words 'the imperfect reproduction alerts you to features of the painting that you might otherwise have overlooked'. This suggests that one is better placed to identify safe behaviour if one has seen a range of behaviours that might be described as unsafe (and vice versa). Certainly, the research data suggests that some individuals are making that type of comparison in their thinking about safety. Many interviewees at the chemical plant and the nuclear power station exhibited a very personal awareness of the real dangers involved in working with hazardous technologies by spontaneously recounting stories (often dating back many years) where some unexpected event had frightened them enormously.

The power station had experienced only one significant incident in its operating history that was serious enough to lead to unfavourable publicity and also a fine from the regulator. One of the operational managers was working on shift in a more junior capacity when the incident occurred. He still remembers details of that shift, such as the date and day of the week. He said, 'I think you've got to go through some experience like that … shall we say you've been blooded then and you realize that sometimes decisions you make thereafter, what implications can come from those decisions or not taking decisions.' One operations manager at the chemical plant told a similar story about an incident he was involved with:

> I was involved here maybe five years ago now, where we had an incident which was potentially quite hazardous … [it] is certainly something I will never forget. It was quite, in a lot of ways it was quite traumatic from a personal perspective, from my own personal perspective anyway. … I think at the time your training kicks in and you just think about making the plant safe. Afterwards you just think what could have happened … and you think what happened at Longford.

Those are the sort of things that go through your mind particularly afterwards when the dust has settled.[1]

There are fewer stories of this kind in the research data from the air navigation service provider, but this is perhaps due to the method of data collection (workplace observation and somewhat disjointed conversations in an operational environment, rather than one-on-one reflective interviews away from the workplace). Nevertheless, Story 26 is a similar case, where a story about an accident that he was involved in many years earlier was what came to mind for one operational manager in his desire to explain why conservatism in safety decision-making is so important in an air navigation environment.

Also, two of the operational managers at the air navigation service provider independently mentioned the Lake Constance accident[2] as a factor in their thinking about safety issues. This European experience is the only major aviation accident of recent years that has occurred substantially due to air traffic control system problems. The operational managers mentioned this incident to make the point that it would be possible for something similar to occur in Australia. This was not meant as a criticism of any specific aspects of the Australian air traffic control system or a generally disparaging remark. It was rather an acknowledgement on the part of the operational managers that the system is not perfect and the potential for serious accidents is always present.

Other operational managers had experienced a similar highly memorable moment in their careers, brought about by a much smaller event. These could be small, even transient, operating anomalies that in some cases were not even noticed by the other operating staff at the time. The key factor was that the system had behaved in a way that the operational manager did not expect and hence from their perspective was temporarily out of control. Weick calls this a cosmology episode. 'A cosmology episode occurs when people suddenly and deeply feel that the universe is no longer a rational, orderly place.' (Weick 1993: 633). McAdams (1993) calls them 'nuclear events'. The key factor is the shift in perception – an understanding of the twin challenges of danger and uncertainty posed by the system. This is the moment that 'the beast' (see Section 6.5) was born in the mind of the operational managers.

---

1   He is referring to the Esso Gippsland Gas Plant fire and explosion at Longford in 1997 in which two people died. See Hopkins, A. 2000. *Lessons from Longford: The Esso Gas Plant Explosion*, Sydney: CCH.

2   In this 2002 incident, two large jet aircraft collided in mid-air over the German town of Überlingen. One aircraft was a charter flight from the Soviet Union with a party of school children on board. The other aircraft was a cargo plane. Everyone on both aircraft died in the incident – 71 fatalities. Many of the root causes of the incident were found to lie within the Swiss air traffic control system – see BFU 2004. Investigation Report AX001-1-2/02. Braunschweig: German Federal Bureau of Aircraft Accident Investigation.

The idea that you must experience some traumatic event to be part of the professional elite is a common one. In the BBC documentary program about the Lake Constance accident, one of the air traffic controllers also describes what he believes to be the experience necessary to truly become a professional: 'When you start to work as an air traffic controller it's fun, it's great. It's all very easy, it's all very good and then at a certain stage you come across a situation where there's a near miss or you're in a situation where suddenly your confidence in yourself is a bit shaken because you've done something or there was an error that could have led to a difficult situation.' (BBC 2003)

The operational managers have risen to the highest level of their respective professions over an extended period of time. Many of the incidents that they spontaneously recall when thinking about influences on their decision-making are things that occurred many years ago. Their first-hand experience of key events in the history of each site is another benefit to the organizations of their continuous and long term involvement. Several operational managers in different organizations mentioned the cyclic nature of safety awareness within their organizations. They had been there long enough to see the organizations go through cycles of complacency, near-disaster and sustained effort. In some cases, they were concerned that they may be starting to see early signs that their organizations may be moving from effort to complacency. Ironically, this concern makes it less likely that such a shift will actually occur. As described in Section 6.5, they view the system with which they work as 'the beast' – something complex, dangerous and malevolent that needs to be dynamically controlled for the system to be safe. The implication of the stories recounted here is that the operational managers have a sense that the beast is ever-present and ready to break out. Pariès (2011) highlights such an awareness as a necessary quality of resilience – that not all events can be anticipated and hence operational experts must be 'prepared to be unprepared'.

The ability of operational managers to see the difference between safe and unsafe requires that they have experienced what both situations feel like. This rich experience makes for better decision-making and is certainly consistent with the view that people in high reliability organizations are very aware of the possibility of failure (see Section 2.1). Their understanding of what it means to be safe or unsafe is context-specific, practical and intuitive rather than generalized, theoretical and analytical. This is consistent with Dreyfus's (1986) work on expertise and learning (see Section 2.5).

Whilst awareness of danger is important, being hyper-aware of the dangers posed by the system, or not confident of one's ability to keep the beast caged, can inhibit decision-making, as described in the next section.

8.1.2 Being in control

The other key factor in identifying cues is the feeling of being sufficiently in control. Developing problems can go unnoticed, as they are literally unthinkable without the skills to deal with the implications. Less experienced people may fail to see a problem, not only because they lack the technical ability to interpret the situation, but also because they lack the capacity to deal with it and hence are psychologically incapable of acknowledging that the problem exists.

This inability to see, despite significant evidence, has been described by Westrum (1994). He published a case study demonstrating the power of what he calls schemas or what naturalistic decision-making researchers (see Section 2.6) might call mental models. A schema or mental model is a set of relationships describing how ideas and events are interconnected in the form of cause and effect. It allows us to make sense of day-to-day occurrences, seeing how each small part of our existence fits in to both the past and the future. Westrum's case study involves radiologists in the USA in the 1950s and 1960s. With improvements to x-ray technology and hence its increased use in patient treatment, several radiologists became aware of the significant number of children presenting at hospital showing signs of broken and healed fractures in the long bones of their arms and legs. The parents of these children were unable to give any explanation of the injuries. Medical staff began to suspect that perhaps some kind of undiagnosed condition had made some children's bones particularly brittle and hence vulnerable to fracture from relatively trivial impacts. This condition was given the name 'unsuspected trauma syndrome' and the causes remained a mystery. The breakthrough came when several hospitals established multi-disciplinary teams including radiologists, paediatricians and social workers. The social workers were used to dealing with issues of violent and dysfunctional families. Together, these teams soon realized that they were dealing with a serious social, rather than medical, problem. Soon afterwards, a very influential paper on their findings was published in the Journal of the American Medical Association. The observed condition was renamed 'battered child syndrome'.

In Westrum's view, the medical profession could not make the connection between the evidence of the children's injuries and the increasingly obvious cause. The fact that parents might be harming their own children was literally unthinkable as it challenged the schema of the doctors regarding social relationships and behaviours. A change to their schema became possible only once they had a way of addressing the issue (by linking with a different group of professionals who had the vocabulary and the skills to address the problem, that is, the social workers). The generalized version of this observation is that 'the system cannot think about that over which it has no control'. (Westrum 1994: 336)

This example has been cited on several occasions in relation to sensemaking and extracting cues (Weick 1995, 2006) and is directly relevant to safety decision-making in an operational setting. 'When people develop the capacity to act on something, then they can afford to see it. More generally, when people expand their repertoire, they improve their alertness. And when they see more, they are in a better position to spot weak signals which suggest that an issue is turning into a problem which might well turn into a crisis if not contained.' (Weick 2006: 1724)

This suggests that the operational managers are capable of seeing cues of abnormal behaviour because they believe they have the skills to address them.

Importantly, the incidents described in Section 8.1.1, whilst frightening in their own right, were all near misses from the perspective of the interviewees. None of the operational managers mentioned any incidents that they had been involved with that resulted in fatalities. They had been shocked by the potential, rather than actual, personal impact of what might go wrong. Information about the Lake Constance accident also shows how important it is for those involved with high hazard technologies to believe that such situations are controllable. Immediately after one of their colleagues made an error that led to 71 deaths, the potential consequences of their actions seem outside their control. Interviewed after the accident, one controller said 'Until that day, we always had this safety net. Whatever happens, they won't crash in to each other with today's technology, but that safety net has been taken away from us.'[3] (BBC 2003) And another controller: 'If I think that the small blip on my radar screen is an aircraft with 300 people, 300 persons on board then I can't do my job. I have to stop. I have to say goodbye and get another job.' As a third controller said: 'To see suddenly what is the destructive power of our job. If one imagines, if I make an error and it results in 71 deaths, I don't know if you get up in the morning and go to work happily and calmly. Sometimes you need to be calm and cool to do the job.' (BBC 2003)

Whilst several of the operational managers at the air navigation service provider talked about the Lake Constance event, there was another consequence of the event that none of the operational managers mentioned. The air traffic controller whose error was one of the immediate causes of the accident was murdered by a Russian man who had three family members killed in the accident. It could be argued that the failure to mention this outcome (that they would all have been well aware of) was because it is outside their control. The operational managers have the ability to influence technical aspects of the system, but appalling outcomes such as this are completely outside their control, and hence unmentionable.

Published research and accident responses such as those cited above provide a warning that operations staff can be overwhelmed by their experiences if they feel that danger is outside their control. In contrast, the field data suggests that the operational managers at the nuclear power station, the chemical plant or the air navigation service provider, whilst being well aware of the potential dangers they are responsible for, all feel confident that they have the ability to control any surprises that arise.

HRO research (Weick et al. 1999) suggests that preoccupation with failure is one quality of a high reliability organization. In one sense, this quality of HROs is the opposite of a complacent attitude of thinking 'it can't happen to us'. The operational managers interviewed were very well aware of the potential for failure and the seriousness of the consequences for themselves, their colleagues, the public and their organizations. The challenge with safety is to ensure that the

---

3   The safety net the controller is referring to is a cockpit alarm system (known as TCAS) that warns a pilot of other aircraft in dangerous proximity and advises what action to take. In this case, the TCAS system failed to prevent the accident.

anxiety produced by such events is translated into concrete useful action, rather than an unhelpful emotional response, such as repression and denial.

8.2 Applying Experience to Decision-making

Operational managers are able to see developing safety issues because they have the experience of both safe and unsafe situations, plus the confidence that they are able to control developing issues. This section addresses another way in which operational managers use their experience to decide what course of action to take in any given situation. Classical decision-making tells us that the preferred way to proceed would be to develop a (mental) list of options and evaluate each one against a set of appropriate criteria. In fact, operational managers use much less analytical and more personal ways of deciding if their proposed course of action is a good one. There were three types of 'tests' described during interviews. Each one involves an operational manager imagining himself to be at a point in the future in a situation that makes him vulnerable to the possible consequences of his decision.

In the first type of test, some participants talked of imagining that they were taking a family member into the area that could be impacted if something were to go wrong. If that made them feel uncomfortable, then they knew that the situation was unsafe. One operational manager at the chemical plant described his decision-making as follows:

> I don't make an environment that I wouldn't be comfortable for one of my children to come in to. I used to use my brother, but now I've got kids it's easier to use my children. Now I accept that I have a four year old so you wouldn't bring him onto a plant necessarily, but it's the same concept. If I wouldn't carry him around and show him all the things and feel that he was safe, then why would I allow the guys to work in that same environment?

The second (more common) test was about the personal consequences if there was an accident. No-one expressed fears for their own personal safety, but several operational managers expressed significant trepidation at the idea of having to tell the family of one of their work colleagues that a family member had been injured or killed as a result of a decision that they had made.

> I have unfortunately a couple of times in my career had to go to hospital and had to talk to someone who is in hospital and then go and speak to their family members ... luckily they weren't really horrible ones – not a fatality for example ... and that also frames some of my decision-making ... How would I feel going and explaining that series of decisions to somebody's family?

The following quote from another chemical plant operational manager shows a similar fear for his own emotional safety.

> Many, many years ago I had a pipe fall and crush my finger and the worst thing my boss had to do, ... he had to phone my wife and say I'm very sorry your husband is going to hospital. Apparently he broke down when he had to do it. I always dread having to phone somebody and say your husband has gone to hospital. ... I know [my current boss] would probably have to make the phone call but still, having been there, I don't want that.

These comments were made in response to questions about how they decide what to do in any given operational situation. Imagining the personal consequences of dealing with the potential bad outcomes was a link made by these interviewees.

The third type of personal test is the 'newspaper test'. In this case, an operational manager considers how he would feel if his decision were published for all to see on the front page of the newspaper. In this case, the interviewee was an operational supervisor at a large refinery operated by a major oil company who was describing how risk management worked in practice in his organization. Whilst working for a different organization, the position held by this person was equivalent to those held by the interviewees in the organizations whose decision-making practices we have explored in detail. Early in the interview, he explained in detail the intricacies of a new risk matrix used across the corporation. He felt that this was a great improvement on previous risk matrices that he had used as it went into more detail in classifying risks and made the overall process of assessing operational risks more objective.

Later in the same interview, he was asked to describe what factors he considered in deciding whether to sign off on risk scenarios that were presented to him for approval. His organizational position was such that risks identified as being at the second highest level on the corporate scale, plus any plans to further reduce the risks, needed to be approved by him for the part of the plant that he worked in. The list of factors he gave was:

- His gut feel about the scenario – is it credible?
- Who was involved in the risk assessment? Does he trust their opinions?
- Are the proposed actions really going to address the issues raised (not just raising paperwork, but actually taking action in the field)?
- What do others think about this issue (his boss, his colleagues)?
- The newspaper test – how would I feel if my decision were to appear on the front page of the newspaper?

We see here that whilst the formal risk management procedures on the site determined which scenarios were presented to this person for approval, once they had been brought to his attention, his decision-making was based on the sense he made of the information before him. Ultimately, his emotional response to the situation based on his past experience was as important as the analytical, risk-based conclusions of the initial assessment by less expert individuals.

In all three types of mental tests, the operational manager is concerned about his own mental safety. The potential for powerful negative emotions such as grief, guilt and shame provides the incentive to make the right choice. These tests of possible courses of action by imagining the consequences are similar to the process of recognition-primed decisions described in Section 2.6. This research data suggests, however, that in the cases of high hazard organizations the mental simulation used by operational managers is about the potential emotional, rather than just technical, impact of their actions.

8.3 Stories and Learning

As discussed earlier, stories play a key role for operational managers in their thinking about safety. This includes stories about past accidents and incidents that give a sense of what it means to be safe or unsafe (see Section 8.1.1), and stories that provide a way of exploring how they feel about specific courses of action they may choose (see Section 8.2). But stories have yet another place in operational decision-making. Storytelling is an important means of continuing professional learning for the operational managers – increasing their technical knowledge of the behaviour of the system, as well as having emotional value in dealing with working in a dangerous industry.

Sullivan (2005) links the Dreyfus and Dreyfus model of expertise described in Section 2.5 with professional learning. As he points out, the Dreyfus and Dreyfus model reverses other models of adult learning – moving from context-free rules and general principles to learning from practice and practical experience. In this model of learning, practice drives theory, not the other way around. Accepting this model of adult learning, then, it is not surprising that operational managers in all three organizations love to share stories of their experiences in working with the system. These are the stories of the unexpected things that happen – usually not stories of (near) catastrophe, but technical details of unexpected system behaviours. Such stories have relevance only to those professionals who already have a deep technical understanding.

It might be expected that a comprehensive incident reporting system such as those seen at each of the three organizations studied might provide interesting and useful feedback to operational managers about system behaviour. In fact, no interviewee (at any site) mentioned the formal incident reporting system in the context of sources of operational learning, apparently because these systems have not been designed with this type of learning in mind.

The importance of incident reporting systems has increased over the past decade in high hazard organizations – even those where people are not specifically aware of HRO research emphasizing the value of learning from minor incidents. Attempts to apply error management (Rasmussen 1982) and organization error

(Reason 1997) approaches to safety have led to incident reporting systems based around classification of error types and contributing factors in order to allow trend analysis and develop remedial measures. In this research, each of the participating organizations is aware that collecting data about incidents is important for safety performance improvement. Each therefore has in place a comprehensive system of incident reporting.

Broadly speaking, the systems are similar. Everyone in the organization is encouraged to report incidents and near misses. Operational managers receive copies of reports so that they can act immediately on operational issues. All reports are followed up by the safety department to ensure that appropriate corrective actions are put in place. Safety department personnel also undertake some degree of trend analysis in order to extract general lessons about potential problems. At best, these systems provide an important feedback loop to senior management about how their decisions and actions may be impacting the safety performance of the organization as a whole and also identify many system changes that can improve safety performance. The most interesting thing about these systems in relation to operational decision-making was how few links there seemed to be.

The power station definition of things that should be reported via the incident reporting system covers 'any actual unplanned deviation from normal operating conditions, procedures or practices that results in loss such as injury, occupational illness, fire or explosion, property damage or plant trips costing over £500, environmental discharges or any near miss events that could have potentially resulted in such a loss'. The emphasis is on loss – real or potential.

In practice, this is resulting in a wide range of events being reported (averaging over 2000 events per year and rising) (Hayes 2009). In one sample week, reports included:

- equipment failures with possible safety implications (such as failure of a cooling water pump);
- procedural breaches with possible safety implications (such as a fire door left open);
- incidents in which someone was (or could have been) injured or exposed to radiation (such as a routine radiological survey showing very low level contamination in recreation room);
- observed conditions directly creating a hazard (such as potholes in the approach road that may be a hazard to cyclists); and
- observed shortcomings in safety systems (such as poor coverage of the emergency public address system in one area).

Given that approximately 650 people work at this site, the number of reports averages more than three per person per year. This is impressive and equals the best reported rate for the chemical industry (Kjellen 2000, Phimister et al. 2003).

The name of the system (Operational Experience Feedback) gives a strong indication of the way the designers intended it to be used. Many individuals and departments take advantage of the vast store of data contained in the system in various ways. Each plant operational shift and each maintenance group has at least one nominated Operational Experience Feedback Communicator. These individuals generally act as liaison between the Safety Department and their specific work groups on these issues, ensuring that their co-workers are aware of any new reports, investigation findings or actions that are especially relevant. They are experts on finding their way around the large amount of data available via the company intranet and helping their workmates find any information in the system that is relevant to specific tasks at hand. Use of the data in a real time operational sense for work planning was at an embryonic stage at the time the research fieldwork was done, but champions of the system in the operational teams could see the potential, and word was spreading.

Operational managers supported the use of the system in this way, but it was of little direct benefit to them in their own work. The exception was the very small proportion of reported incidents that related specifically to unexpected operating anomalies with the reactors and associated plant. These would normally be reported by the shift that was on duty when the issue arose. Personnel on other shifts (especially Shift Managers and other control room personnel) are very interested to find out what has occurred. In especially interesting cases, the details were referred to the simulator engineer, who can set up a new case on the simulator of the reactor control scheme so that people can experience in real time how the scenario developed and experiment in a safe way with various possible responses.

In a similar vein, the air navigation service provider has a comprehensive incident reporting system, as described in Section 5.3. The focus is on reporting errors made by controllers or others active in the broader airways systems (such as pilots). Operational managers are involved in various ways in this system – managing the implications of serious errors, ensuring reports are submitted, conducting investigations – but air traffic control system behaviour issues are not typically reported in to this system unless the outcome is significant. The scale of activities at the chemical plant means that the incident reporting system is also smaller scale, but the same general processes apply. In a similar way, operational managers are interested in and involved in the incident reporting system, but plant operational quirks and abnormalities would not be reported unless there were (almost) serious consequences.

Although operational managers were intimately involved with site incident reporting systems on a day-to-day basis, the type of incident report encouraged/ produced by these systems was not valued by interviewees as a source of new information to enrich their mental model. There appear to be two reasons for this. Firstly, the information about each incident is not in a form they find useful. The forms (paper or on-line) used to collect incident reports are data driven, dominated by tick boxes and closed questions. Of course those who report incidents are invited to summarize the sequence of events, but the emphasis is on incident classification, rather than collecting a rich or nuanced picture. Rooksby et al. (2007) have made the same point in the medical sector, looking at incident reporting in anaesthesia.

They point out that 'the reporting schemes are not just there to collect data but to afford the stories of what went wrong'. Their work has shown that reporting systems in anaesthesia are often weak in supporting this second objective. Whilst incident reporting was not investigated in detail as part of this research, the same issue seems to apply here.

The second reason that operational managers who are keen to sharpen their mental models did not find the incident reporting systems particularly useful is that the incidents of most interest to them for this purpose were not reported. Unusual operating occurrences are not recorded in the incident reporting system unless there was (or was potential for) a significant loss or failure. Many of the incidents of interest to them would not be understood or noticed by other staff with less experience and knowledge of the operating system. They are of interest because of what they reveal about the system itself, not because of the potential safety consequences per se. These criteria are quite different and, at all sites, operations managers loved sharing their 'war stories' of what had happened on earlier shifts.

At the nuclear power station, the operating staff had even developed their own recording system for these types of incidents, as described in Section 3.3. This was simply a folder in the operations managers' shared office where a written summary of each event was kept. Unlike the incident report form, which included many tick boxes and classification questions, the form used for recording purposes was essentially free text. These records were highly valued by the operations managers and regularly referred to. In the other two organizations, there was no system for recording stories of operational quirks.

In all three organizations, stories of recent operational quirks were eagerly shared in informal conversations at shift handovers or on other occasions where two or more operational managers found themselves together. Stories were sometimes passed on by email. This type of communication was valued by all, despite the practical difficulties presented by their shift arrangements.

In summary, operational managers improve their understanding of the technical systems with which they work (and hence their decision-making) by their own experiences and sharing the experiences of others in the form of stories. Formal organizational systems for incident reporting are of little use to them in this context because the information collected is not in the appropriate form, and also because many of the stories that are of interest are not defined as 'incidents'.

8.4 Summary and Concluding Comments

Operational managers are highly experienced individuals with a deep and specific understanding of the system within which they work. Their knowledge and understanding of the system are story-based in three different ways:

- They remember stories of past situations that they believe to be unsafe, and use these for comparison with new situations to make a judgement about what is safe and what is not. These stories may be linked to their role in a serious incident, or they may be something much more minor but still unexpected, that frightened them early in their career. These stories reinforce the idea of the system as a beast that is dangerous, unpredictable and needs to be controlled.

- Operational managers also imagine possible future events in the form of stories when considering possible actions. Again, these are not stories based on technical details but stories in which they imagine their own role and their feelings in taking up that role – 'how would I feel if … ?'
- Stories are also an important way of expanding their technical understanding of the behaviour of the system. Operational managers love to share stories of their experiences and those of others about unusual things that have happened and how others chose to respond in particular cases.

Stories are valued and shared within their professional group as a way of increasing specialist knowledge and expertise. The specialist nature of these stories and the form of the information (rich description, rather than data driven) mean that there are only weak links to broader organizational incident reporting systems.

In any given situation, sensemaking is a retrospective process, linking cues in the present to memories of the past, and stories – told and repeated – are fundamental to retaining experience and making it accessible in the heat of the moment. Schank (1990) describes this: 'We need to tell someone else a story that describes our experience, because the process of creating the story also creates the memory structure that will contain the gist of the story for the rest of our lives.'

Operational managers apply their experience and judgement to create an environment in which production can continue safely. They do this by maintaining a broad view of operations at any given moment and anticipating problems before they arise. They are at their most successful when they seem to have least to do. To an uninformed observer, the lack of activity can be taken to mean a lack of complexity and reduced need for monitoring of the system. In reality, it may mean that the operational manager has successfully enacted an environment that, for the moment, supports all organization goals – safety and production. A high level of activity may be a sign that something unexpected has occurred and that safety margins are threatened, rather than a sign of productivity.

Chapter 9
Professional Relationships

So far, the discussion has focused largely on decision-making by individuals, but sensemaking is fundamentally a social process. It follows, therefore, that decision-making also has a social dimension even when decisions are nominally made by individuals.

This chapter turns firstly to the relationships within the operating team (Section 9.1). Leadership is about creating an effective environment for decision-making. Operational managers tell stories of their role in relation to their subordinates that focus on using their experience and judgement to help team members see what is important in a given situation. Section 9.2 turns to the relationship between operational managers and their peers and the key role those relationships play in operational decision-making. The meaning we attach to environmental cues is influenced by the real or even imagined presence of others. We see this in all three case studies where operational managers seek to consult their peers – partly no doubt due to their technical expertise but also significantly because of the emotional support that comes from consulting professional colleagues about important choices.

Another important relationship is that between the operational managers and more senior management. The key factor is the level of trust. There is a high degree of uncertainty in safety decision-making (even after the event), and so it is important to operational managers that their judgement is trusted by the organization at large. This is explored in Section 9.3.

9.1 Relationship to Subordinates

The operational managers have a high degree of confidence in their operating crew to manage the details of abnormal situations when they occur. As we shall see, the operational managers therefore focus their energies on three things:

- Managing the system as a whole to try to prevent operating issues from arising and putting pressure on the team,
- In the event that operating problems occur, observing the activities of the operating personnel to ensure that they are responding appropriately, and
- Managing the broader implications of operational problems whilst leaving the details to others.

In all cases, the operational manager works as leader of an operations team. Operational managers have a fixed team of people working for them at both the nuclear power station (22 people) and the chemical plant (five people). The group stays together whether they are working day shift, afternoon shift or night shift. At the air navigation service provider, at any specific time the operational manager is head of a working group of people, but the composition of the team changes as rostering is done at an individual rather than a team level. Nevertheless, the total pool of people is not large and the operational managers know all possible team members well.

The general attitude of the operational managers to their subordinates seemed to be one of significant respect. No interviews were conducted with subordinates, but there was no suggestion in any of the workplace observations of any general difficulties in this relationship from subordinates. One of the nuclear power station managers (interviewee 1) cited the experience of his shift as one of the two most important factors in making good safety decisions.[1] He felt that their combined level of experience meant he had a lot of 'expert advice available on which he can draw when things go wrong'. His phraseology indicates the value he places on the input of his subordinates, but also his understanding that he retains the overall responsibility for decision-making.

One of the interviewees at the nuclear power station described at length how he interacts with control room staff and how his role differs from theirs. This was possibly something he had thought about more than others, as he holds a licence for some control room positions and sometimes does relieving shifts at that level. In that situation, he is working under another operational manager and he was able to articulate quite clearly how the roles and responsibilities for operational decision-making vary between these two positions. Some of the operational managers at the air navigation service provider were also in this situation of doing relieving shifts in a lower position and they, too, volunteered more information than their colleagues about the differences between the roles.

The nuclear power station manager described the operational response to an abnormal event as having three levels. The levels take a successively higher-level view of the situation and necessary actions and look further into the future. As he explained: 'There's no point in three of us looking at the same switch. Whilst one guy is turning it, two guys are saying yes, it's definitely going lower, that's good. Meanwhile behind you there's a fire going.'[2]

In this manager's thinking, the panel operator works at the most immediate, or base, level – literally opening and closing valves, starting and stopping equipment and changing control set points in response to the situation. The control room supervisor checks what the panel operator is doing and considers what the next step should be. The manager described this as 'base plus one'. In contrast, the operational manager works at 'base plus two'. He needs to be thinking conceptually

---

1   The second factor was support from senior management.
2   This is a metaphorical, not literal, fire.

about the overall plan to resolve the situation and what might be the result of each possible course of action. The interviewee drew a diagram that had the form of a tree, emphasizing that there could be 'hundreds and hundreds of branches' that he needs to consider in coming to the best overall plan.

Having said that, the operational manager still took responsibility for ensuring that the base and base plus one levels of activity were going on. He described one specific example:

> A couple of years ago, a turbine went on us [was stopped automatically by monitoring instrumentation due to an operational problem]. The desk engineer was quite experienced but he had never had that situation and to be honest he struggles a bit. You have three types of people. People who will freeze and do absolutely nothing, people who will push every button going. They haven't got a clue but they'll get the right one eventually maybe. And the type of people who will be measured. That's the three types we see … That's another thing I try to do. If the guy is spinning and doesn't know what's going on, now the supervisor has to do all the work. You might as well wheel him out of the way. If you can give him a little bit of confidence and watch what he's doing, he might do nine out of ten things. Then all I have to do is think about one. Now I'm thinking, aren't I, instead of doing his job?

Another operational manager at the nuclear power station recounted a story of a water leak that required quick repairs. In recounting the story of what transpired he said: 'One or two times I thought I'm getting too close to this. I need to stand back and have the bigger picture of what's going on and manage the situation rather than me being in there prodding and poking to try to get the repair sorted out.'

One of the operational managers at the chemical plant (interviewee 3) also gave a detailed account of a case where the course of action he chose to follow was determined partly by the level of experience of the control room operator on duty at the time. In this case, he had used the 'line in the sand' approach. The window he gave himself and the rest of the crew to fix the problem was based on his confidence that the control room operator could be relied upon to intervene and shut the plant down if the situation deteriorated further and reached the limit they had decided on. In fact, things did get worse and the control room operator took the necessary actions to shut down the plant. The operational manager indicated that he would have made this decision earlier (and so chosen a different line in the sand) if the control room operator on duty that day had been less experienced.

A similar hierarchy within the operating team exists with the air navigation service provider. Several situations were observed when a developing abnormal event led to a three level response rather similar to the levels described by the nuclear power station manager. In each case, the general distribution of duties was:

- The air traffic controller in contact with the traffic involved directly in the situation maintained a detailed level of monitoring and response,

- The aisle supervisor ensured that related detailed operational matters were covered (for example, by reallocating tasks to other controllers), and
- The operational manager planned the next likely steps in the required response (for example, callouts of other agencies, or the need for extra staff or changes to traffic).

In a similar vein, one of the Sydney traffic managers described his job as 'manoeuvring to keep the ship on course' rather than doing the controller's job for him. Another expressed his pro-active management of the system by saying, 'It's like a frog in a pot of hot water.[3] The problems can sneak up on you [when you are a controller]. My job is to keep an eye on traffic levels and make sure this doesn't happen.'

Theoretical discussions of sensemaking emphasize that making sense of a given situation can mean literally creating a new environment. Operational managers take on this leadership role for their operating team. They seek to create an environment in which the focus is on matters the managers believe to be most important so that the potential for further problems is anticipated and the entire operating team can work calmly and safely.

9.2 Consultation with Peers

Since the operational managers work on shift and in all cases the organizations run 24 hours per day, seven days per week operations, the operational managers never meet as a group. Opportunities for face-to-face interactions with peers are limited to shift handovers or *ad hoc* conversations when one manager is on a special day shift assignment (for example, providing operational input to a major project). In a study of airline executives and middle managers conducted in the early 1990s Mouden (quoted in Westrum and Adamski (1999)) determined that managers at all levels found communication on safety issues with their peers, more effective than along vertical lines in accordance with published organization charts. The behaviour of the operational managers also shows that they value the input and opinions of their peers. As described above, the individual operational managers are, without exception, very experienced in the technology and facilities for which they are responsible. They were also very clear in most cases that responsibility for operational decision-making lies with them. Despite this, most managers took the opportunity to consult their peers before making a decision if the circumstances allowed.

---

3   This refers to the 'boiling frog' metaphor whereby a frog resting in a container of tepid water that is slowly heated is fooled by the slow temperature rise and dies rather than jumping out of the container. The metaphor refers to a lack of ability to perceive the incremental impact of small changes.

At all three sites, interviewees know their peers well. They have a long history of working together – in many cases their entire working careers. This brought to the workplace a sense of stability that contrasts strongly with many other organizations (and, in some cases, other parts of the organizations studied). Although contact between operational managers is quite limited, many of them have worked together on shift earlier in their careers and they all know each other very well.

Both at the nuclear power station and at the chemical plant, several of the operational managers had spent their entire career at the site (through several changes of ownership). At least two of them worked together on commissioning the first stage of the chemical plant more than 27 years previously. Also, two had left the organization to work at other chemical plants in the area but had chosen to return after only short periods away – 'I saw the light and came back'. Several of the operational managers at the air navigation service provider joked about how long they had worked together – they were in the same graduating class of air traffic controllers from the in-house college – the Class of '82.

This fosters relationships within the group and a strong sense of trust in others' experience. In all three organizations, operational managers typically consulted their colleagues (if time allowed) when making significant operational decisions. This was most apparent at the nuclear power station. Like all major industrial facilities of this type, the site has a formalized system of emergency response and after hours call out. Part of this system involves an emergency team made up of senior site specialists who have been trained to fill specific roles in the event of an emergency. Senior staff are rostered to fill slots on the team and carry pagers for those weeks when they are rostered on. In theory, they are on 24-hour call for site emergencies. In practice, however, an informal site practice had grown up where the on duty operational manager was likely to ring members of the emergency team for advice in the event of operational upsets outside normal office hours which in no way constituted a site emergency.

It was most common for calls to be made to the duty emergency controller and/or the duty reactor physicist. The reasons for calling people holding these two roles differ. Most of the duty emergency controllers are ex-operational managers who have moved into office-based positions. The reasons for calling operational colleagues were described at some length by interviewee 8, one of the people who fills the duty emergency controller position:

> Out of hours you are lonely on shift and you've got the SOIs [Station Operating Instructions] but usually these kind of events don't fit in with procedures. It's always something different every time. I've done it myself when I was [operational manager]. You get all sorts of funny things and you ring controllers up at two and three in the morning. You just want to test your argument with someone else. Anybody would do, but the controller is on standby and he's getting paid for it. What they want is a sounding board, just to test their arguments I think. Just the confidence to say when they come in in the morning

well I spoke to blah de blah and they didn't disagree either so I went ahead and did it. At the end of the day, I say to them, you're the DAP [Duly Authorised Person], you make that decision. Really they could have done it without ringing me. But I've been in that position on shift myself, it's just ... I don't know ... it's a difficult call sometimes because you've got 'well I could keep it on but something in my mind is telling me to take it off' kind of thing. I've been in that position with the reactor when I was on shift ... You're lonely at night or out of hours ... That's the reason they ring me up I think and I don't mind talking it through with people when they ring me or getting other people involved.

The interviewee's comment that people are paid to give this type of advice is actually incorrect. There was some minor confusion at the site over the issue of the role of people on the emergency duty roster. A proposal was under consideration to train other non-operational managers in the requirements of the role of duty emergency controller and include them on the roster. Those putting in place the new arrangements were unaware that the much more frequent call on the services of the duty emergency controller was to provide operational advice, something the new inductees to the role were not at all qualified to give. In practice, some on duty operational managers would call the person they were most comfortable with to ask for advice, rather than the person on emergency duty, and it is likely that this would continue in the future.

Some out-of-hours calls were also made to peers in a different but related profession. The reason for calling the reactor physicists (or occasionally other professional specialists) was typically technical advice rather than reassurance. On occasion, events occur that are outside the technical understanding of operations staff, even at the most senior level, and they have no hesitation in seeking specialist advice. An example of this is given in Story 6.

Similar consultation arrangements apply to abnormal operating situations that arise during the day. Duty emergency staff are usually on site during business hours and other senior staff may be available for consultation. Details of all advisory conversations are recorded on special forms, as described in Section 3.3, although the explanation given by the operational managers about the forms themselves relates to documentation of decisions for potential legal reasons.

Operational managers at the chemical plant also consult their colleagues on many occasions. They too have an emergency callout system, but the processes for getting out of hours advice are not linked to this system. They are managed much more informally. Many stories were recounted where one manager had called either one of his colleagues or someone from the maintenance department, for advice outside normal working hours (see, for example, Story 14). There was some suggestion that professional pride might be preventing some of the younger and relatively less experienced operational managers from calling their senior colleagues for advice until they had exhausted all options that they could think of for themselves. During normal business hours, there are typically two other individuals with the same level of experience as the on-duty operational manager available at the site (due to day shift assignments). Many decisions involved all three managers, but they were also clear that the person on shift was the person with the final responsibility (see Story 16).

Consultation between peers was also very obvious at the air navigation service provider. A large part of the job of the operational managers was to consult various stakeholders (managers at other locations, airlines, maintenance department, aviation users) and reach the most appropriate decision. Given the high level of consultation that goes on during any shift as part of their normal workload, there is no clear increase in consultation when abnormal events occur. There was one case described when a developing problem overnight led one operational manager to call his colleague in the middle of the night and ask him to come in early. This was accepted as a reasonable request. The face-to-face interactions between operational managers in resolving operational questions were very similar to those at the chemical plant. Story 16 and Story 24 were very similar in respect of the patterns of interactions between the operational managers in the two cases.

Given the shift work nature of their jobs, it was essentially unheard of for the operational managers in any of the organizations studied to meet together as a group. Individual pairs of managers met regularly as part of shift handover and occasionally in other circumstances. Despite this, operational managers knew each other well as a result of their long professional involvement and were keen to consult their peers in decision-making when time allowed. They had developed an informal network for doing this and were generally supportive of receiving calls from colleagues at any time of the day or night. These calls were generally seeking a sounding board – reassurance that the operational manager's planned course of action was appropriate.

9.3 Trust of Senior Management

As described in Section 6.2, operational managers hold ultimate operating authority, which puts them at the top of the professional hierarchy. This is in contrast to their position in the managerial or bureaucratic hierarchy where all operational managers are in upper middle management, reporting to organizational superiors. It is therefore important for operational managers to feel they have the trust of their organizational superiors in decisions that they take to interrupt operations so as to avoid any conflict between these two aspects of their role.

Many interviewees at both the nuclear power station and the chemical plant said that a key factor for them in making a decision to interrupt operations was a feeling that they would have senior management support. Knowing that they have that support leaves people free to make what they feel is the best decision based on their judgement of the circumstances. This was mentioned several times by personnel at both sites, often with specific examples given. If the plant needed to be shut down for safety reasons, then the site manager was always informed, after the event or, if time allowed, perhaps beforehand. Even if the call was made in advance all personnel were very clear that it was a call to keep the site manager informed, rather than a case of seeking approval or even endorsement.

This nuclear power station operational manager could have been speaking for not only his colleagues, but also the operational managers at the chemical plant when he said:

> I'm sure my colleagues all feel the same. I don't feel under any pressure to justify shutting down or going backwards on production. None whatsoever. Even if I made a wrong call, a call that was shown later to be probably not the right one, I'm pretty sure I would get the support ... I do feel that, I'm sure management understand how important that is. If they are seen to give any criticism or ask for undue justification of a decision made in a safe direction, it actually undermines the whole philosophy of it. It's very key to the way we operate.

The importance of the high degree of trust shown by organizational leaders in the operational managers at the nuclear power station and the chemical plant cannot be over emphasized. As stated in the quotation above, operational managers were confident that they would not be criticized even if, with the benefit of hindsight, they were shown to have interrupted production unnecessarily.

One specific story from the nuclear power station shows this attitude in action. Story 1 is so impressive that it is worth reiterating in some detail. A routine inspection of one of the reactors found some unexpected cracking. The operational manager on duty at the time asked the inspection engineers whether they could guarantee that the same fault was not present in the other reactor (which was still on line) and, further, if similar cracks were present, that it was safe to continue to operate. The inspection engineers were unable to provide such assurances, so the operational manager decided to take the second reactor offline. A week later, further engineering work showed that there was no issue with the reactor integrity and the reactor was restarted. The plant manager praised the conservative thinking on the part of the operational manager, even though he had cost the organization one week's production for something that turned out to be not a problem.

The situation at the air navigation service provider was significantly different. The same degree of trust between operational managers and their superiors was not present (see Section 6.3). All those involved understood that the final responsibility for operational decisions lies with the operational managers, but they felt under significant pressure from those above them in the organizational hierarchy, to make choices that would reduce costs. One of the operational managers said that he would never consult his organizational superiors before making an operational decision. He was clear that 'all decisions are made in the Ops Room'. He would also not generally go out of his way to inform his manager about particular decisions made, unless it was something he felt might result in adverse publicity. An example would be an operational decision to limit Sydney operations to one runway during peak hour (perhaps due to weather considerations), which would result in flight delays and hence potentially complaints. In this case, he would not seek approval in advance, but would let his manager know that this had been necessary.

On the other hand, he said that he would get more senior managers involved before the event in decisions that required money to be spent. Given the nature of the operational manager's responsibilities, this essentially means calling in additional staff. Traffic Manager 4 made a similar observation on informing senior management of issues that might cause adverse publicity. He also said that the only operational decisions in which he would involve more senior management would relate to staffing and overtime.

Melbourne operational managers were also very aware of pressure to reduce maintenance costs. Specifically they felt pressured to allow more routine equipment testing work to be done during normal working hours, rather than at night and on weekends (which incurs overtime costs). These issues are not proceduralized, but are left to the judgement of the operational managers about the potential for safety or operational issues posed by the planned work. The pressure comes from their judgements being questioned by more senior management, but the final decision-making is always left to them. Several specific examples of this were given: for example, one operations manager in Melbourne was aware of some major preventative maintenance work on backup power systems that was coming up. He said that the practice in Melbourne was for the work to be done on a Saturday afternoon. At this time, traffic levels would be low, but maintenance people would need to be brought in outside normal working hours at additional cost. In theory, the work should not impact live systems, but his concern is the potential for unplanned interference with operational systems as a result of the work on the backup system. This type of domino effect is not unknown in a general sense, despite the fact that the systems are notionally independent. The other potential problem is that the backup system might be unavailable if called upon to operate in the event that the normal power system fails for any reason whilst the work is being done on the backup. He was aware that the practice at other centres has been to do the work during routine working hours when traffic levels are higher than on the weekend. Whilst this practice had not resulted in any specific problems, he felt that the practice was poor. In the event of a failure during the week, in his opinion getting the system to a safe state has a 'vastly increased chance of a controller slamming two planes together'. He was expecting to be questioned by his manager as to why he would not let the work be done during the week when this practice had been accepted at other locations. This is in significant contrast to the behaviour of the senior management at the nuclear power station described earlier.

The senior managers at these two organizations seem to have different views about what constitutes a good decision. In one case, a good decision is one based on identification of a potential problem. Evidence has been interpreted in the most conservative way and the decision is seen to be a good one, despite later definitive evidence that no physical problem actually existed. In the other case, evidence is interpreted in the most optimistic way based on past outcomes in a similar situation – we didn't have a problem last time, so it will be OK to do the same activity again.

These differences in attitude illustrate why trust is such an important issue for the operational managers.

Of course, the stakes involved in these decisions are high and trust shown by senior managers builds confidence. Confidence in the ability to control the situation is important in decision-making, as described in Section 8.1.2. Trust also implies a common understanding of what constitutes the right course of action in any given situation, despite significant and dynamic uncertainty. At first thought, it might seem that it should be obvious, at least in hindsight, whether or not a given safety-related decision was a good one. In practice this is often far from clear, because outcome is a poor way of evaluating the quality of a decision.

In a general sense, a poor decision does not always result in a poor outcome. The only clear cut example of this is Story 12, where someone dropped a heavy object from an elevated platform, apparently as a joke. No one was injured, because no one was directly below the area where the item was dropped. This example seemed clear cut to each person who mentioned the story, but the story described above about the need for weekend maintenance on power systems can be seen in similar terms. To the operational manager, the fact that work has been done previously during busy times with no problems is not an indication that this is a good choice. He believes that senior management is incorrect and that this is a poor choice because chance plays a significant role in the safe outcome. As Reason notes: 'The large random component in accident causation means that "safe" organizations can still have bad accidents and "unsafe" organizations can escape them for long periods.' (Reason 1997: 108) In a military context, Janis makes the same point in his work on Groupthink: 'Defective decisions based on misinformation and poor judgement sometimes lead to successful outcomes ... we must acknowledge that chance and stupidity of the enemy can sometimes give a silk-purse ending to a command decision worth less than a sow's ear.' (Janis 1982: 11)

On other occasions, a decision is made to intervene in order to achieve a safe outcome and it is not possible to know what would have occurred without the intervention. This means it can always be argued that a good outcome does not necessarily indicate a good decision (because it might have been an overly conservative interruption). There are many stories like this in the data. Story 4 and Story 19 are both examples where operations were interrupted because of the potential for further problems. No analysis can tell us what would have happened if the intervention had not been made and hence any assessment of the quality of the decision made remains subjective.

The uncertainty about what constitutes a good decision is discussed further in Chapter 8. It is sufficient to say here that the existence of the uncertainty is why operational managers value the trust of their superiors so much. They are making a judgement that in most cases cannot be defended in purely analytical terms. Whilst the logic underlying each individual decision can be explained in hindsight, there will always be issues of judgement around what was known at the time, how it was best interpreted and what might have been if different actions were taken. Operational managers exercise their professional judgement in acting in any given case and they rely on their organizational superiors to reassure them that the choices made do not conflict with what is required of them as employees.

9.4 Summary and Concluding Comments

In making decisions, the operational managers must anticipate the behaviour of the system. There is significant uncertainty about this, even for such experienced individuals. As Reason has famously said: 'If eternal vigilance is the price of liberty, then chronic unease is the price of safety.' (1997: 37) Managing uncertainty in this complex environment is a key driver of the social interactions between the operational managers and their colleagues at various levels. They seek to provide focus and reassure the operating team in times of stress, whilst seeking reassurance in their interpretations from their peers (when time allows), and this uncertainty is also the source of their expressed need for the trust of their organizational superiors. Their superiors generally have much less in-depth knowledge of operations than they do and hence do not have the ability to see all the potential difficulties for themselves, but their organizational position gives them power over the operational managers. The high reliability theorists stress the need to defer to expertise, and this is a good example of that idea. Senior managers rely on the ability of the operational managers to anticipate problems. Expressing a lack of belief or trust in any specific case undermines this significantly.

Chapter 10
Decisions, Risks and Barriers

Previous chapters have shown that decisions by operational managers are based on experience and professional judgement, often within a defined operating envelope. We have seen how a range of professional characteristics such as long experience, trust, integrity and knowledge come together in formulating professional judgements. This chapter turns to the form that those judgements take. Operational managers say that they will not operate unsafely. As one operational manager at the chemical plant put it, '[I ask myself] is it safe or do I stop now?' Those who favour a cultural view of organizational life might say that, in this organization, safety is an espoused value (Schein 1992). To discover what this means in practical terms requires that we look more deeply at the beliefs that inform these declarations about the importance of safety.

Such beliefs are not directly available to the operational managers to articulate, hence their difficulty in providing any generalised account of how they put this concept into practice. Instead, they can be inferred from the actions taken by the operational managers and the stories that they tell. In fact there is a pattern in the stories told by the operational managers about past decisions made. In many cases, the decision about whether a given situation is safe or unsafe appears to hinge on whether existing safety barriers are all in place and available. This use of the concept of barriers was not described in generalised terms, yet it appears to be the structure underlying the actions taken in most cases. Even in cases where the final outcome was not the best possible, the considerations described in each story relate to the integrity or otherwise of barriers of various kinds. Managers' views about barriers are discussed in Section 10.1.

Barriers and risk are related concepts so, in a sense, the decisions made by operational managers are consistent with the overall risk management strategy in each organization. Nevertheless, the operational managers do not use the concept of risk (in the sense of consideration of likelihood and consequence) directly in their decision-making. This comes as a surprise to more senior managers – at least at the air navigation service provider and the chemical plant. Using a risk-based process for safety decision-making, each decision becomes a trade-off between safety and cost (often production). The barriers approach to safety decision-making does not highlight this dichotomy in the same way. This is perhaps why operational managers generally do not see their job as managing the conflict, trade-off or balance between safety and production. They see them as separate goals that can be interdependent but are generally independently achievable. This is important given that much safety theory focuses on sacrificing production in the interests of safety needs. Whilst this theoretical construct may be useful for senior management in the context of risk-based approaches, it does not appear to be consistent with decision-making by operational managers. This is discussed further in Section 10.2.

10.1 Barriers

Of the 26 stories recounted in Part A, almost all involve safety conceptualised in relation to the barriers in place to prevent hazards from leading to accidents, be they physical equipment, procedures or ways of working.

Table 10.1 summarises the stories detailed in Chapter 3. In the vast majority of cases, the need for action was triggered by some change in status of a safety barrier. Barriers include:

- rules defining operating limits (for example process parameters, staffing levels),
- equipment provided for safety reasons,
- production equipment with built-in safety features, and
- operating flexibility/capacity.

The operational managers at the nuclear power station showed the most sensitivity towards changes in the status of safety barriers and hence the most conservative attitude. The stories told show that they would shut down equipment if:

- a safety barrier was lost or partially lost (Story 2, Story 3, Story 8),
- a safety barrier was in danger of being damaged (Story 5), or
- the integrity of a safety barrier was called into question (Story 1, Story 4, Story 6).

Note that in all of these cases it would have been technically possible for the system to continue operating. For example, even in Story 3, which was considered to be a relatively serious incident, it was only the plant monitoring system that failed, not the plant control system. In theory, the plant could have continued to run but the operating crew chose to shut it down since their ability to monitor what was happening (a significant safety barrier in itself) had been lost.

Table 10.1 Nuclear Power Station Story Summary – role of barriers

Story no.	Title	Role of safety barriers
1	Could the same fault be present in the running reactor?	Shutdown due to uncertainty about integrity of one barrier (possible cracked weld in containment barrier).
2	We should have a margin	Shutdown to maintain compliance with operational limits.
3	You need to button your reactors and button your turbines	Shutdown to maintain compliance with operational limits.
4	If another fault develops, the machine might not shut down.	Decreased production due to uncertainty about integrity of one barrier (circulator trip system not functioning).
5	Not a decision you make lightly	Shutdown considered due to potential for damage to barriers from leaking water
6	Show me why it's safe to continue	Shutdown considered due to potential for damage to barriers (broken glass might impair coolant flow).
7	An unusually still day	Extra barriers put in place (gas monitoring and temporary evacuations of some areas) due to loss of one normal barrier – wind.
8	We set boundaries and we work within those	Shutdown considered due to partial malfunction of a barrier (ability to move reactor control rods).
9	Conservative maintenance planning	Avoiding the need to shut down by managing the availability of barriers (in this case cooling water pumps).

Cases where the integrity of a safety barrier was called into question (Stories 3, 4 and 6) are especially conservative. In each of the three examples, a minor fault detected in the system had no immediate impact on any operating safety system. But, in each example, it was identified that the fault could be interpreted as evidence of a latent problem with a safety aspect of the operating system. Rather than simply monitoring the situation for any further sign that a problem might be present or developing, the operational manager took active steps to determine whether a safety barrier was compromised in any way. In two of the three cases, production was curtailed in order to complete the investigation. In this way, potential bad news is treated very seriously and evidence of potential problems is followed up until the exact implications are understood. There is no sense that 'wait and see' is an acceptable option.

Operational managers at the chemical plant also seemed to base their safety judgements in particular cases on potential or actual loss of safety barriers, but a review of the individual stories from this perspective shows how much variation there is in the actions taken, compared with the nuclear power station case. Not only does the nuclear power station have more safety barriers built in, but operational managers there are more consistent in taking action if a barrier is threatened in some way. At the chemical plant, sometimes loss of a safety barrier would lead to a decision to shut down, but on other occasions workarounds or temporary replacements are used to keep the system on line, not always with the desired outcome. Table 10.2 summarises the stories detailed in Chapter 4.

Table 10.2   Chemical Plant Story Summary – role of barriers

Story no.	Title	Role of safety barriers
10	Plant shutdown first as last.	Shutdown immediately to maintain barrier (ie buffer storage).
11	Minimum manning levels.	Partial shutdown (one unit) due to loss of barrier (staff member).
12	A joke gone wrong.	Supervisor dismissed for deliberately bypassing barriers as a joke.
13	Damned if you do.	Shutdown initiated due to potential for barrier failure (stirrer seal). Created a new hazard that was not well managed.
14	A wet weekend.	Shutdown considered due to partial malfunction of a barrier (leaking seal water system).
15	False economy.	Decision made to continue despite partial malfunction of a barrier (reactor cooling water system). Created a larger problem.
16	Just 'til tomorrow.	Decision made to continue despite malfunction of a barrier (inert gas system). Final outcome unknown.
17	Temporary Backup Systems.	Failed barrier (fixed air monitoring) replaced with two temporary systems (temporary loop and hand-held detectors).

Story 11 and Story 13 are both cases where loss of a barrier led to part or all of the chemical plant being shut down. Story 11 is a rare case of application of a rule by the chemical plant operational managers (minimum acceptable manning level). Story 13 is more complex, as there are two different hazards involved. Firstly, due to perceived potential problems with a reactor stirrer seal system, the decision was taken to shut down the reactor. This is a simple case where a threat to the integrity of a safety barrier has led to a shutdown. Problems arose, however, in the response to the shutdown. In this case, the operational crew was left with a partly reacted batch of polymer to dispose of safely. There is a range of safety barriers to deal with the situation and the potential OHS issues that can result. This was managed less conservatively in this case and a more serious problem was created. The operational manager made a direct trade-off. He could have chosen to add chemicals to the partially reacted batch (which would have stopped the reaction but would have required a great deal of flushing to remove from the process equipment). Instead, he chose not to use this safety system (or barrier). This decision led unexpectedly to the reactor being filled with solid polymer, which then needed to be physically jack-hammered out of the reactor vessel whilst managing the exposure of workers to toxic and flammable chemicals. Evidence suggests that an incident like this would not occur at the nuclear power station. In such a situation, the operational manager would have chosen to use the chemical injection system, knowing that it would take time to flush the system. Of course, having chosen that option there would be no way of knowing for sure that the safety and operational problems of physically removing polymer from the system would have occurred otherwise and hence had been avoided.

Story 15 is a similar but less complicated case. Operation continued with a compromised safety barrier (requiring manual control of cooling water) whilst repairs were organized. After 24 hours, control of the system was lost and other safety barriers had to be called into play to stop production. Both safety and production were compromised more than they would have been if operations had been shut down in the first instance until the repairs to the initial barrier were made. On the other hand, Story 14 is a case where a safety barrier (water supply to a seal on a reactor) was damaged and a temporary repair was made, allowing production to continue in the short term so that the repair could be done at a convenient time.

At both the chemical plant and the air navigation service provider, operational managers sometimes chose to make unplanned temporary changes to operations in order to add extra safety barriers if one of the normal barriers was unavailable. Again, this would never be allowed at the nuclear power station. Interviewee 2 at the chemical plant described a case such as this when the fixed air monitoring system (designed to detect leaks from the plant) had a problem (see Story 17). He decided that it was safe to continue to operate the plant with two different temporary arrangements in place, providing what he believed was the necessary safety functionality to run the plant overnight. Night shift passed without incident, but this is a good case to illustrate the point that success must be treated with caution. The temporary arrangements would have been called into service and had their functionality and adequacy tested only if there had been a leak or release of process fluids at the plant overnight. No such leak occurred so it is not possible to know definitively whether the temporary replacement systems were adequate, as they were not tested.

In deciding that the two temporary systems were adequate, the operational manager took into account the specific circumstances by identifying the places most likely to leak and hence locating the temporary barriers where they were most needed. His decision was also based on the fact that there were no further options available. When asked whether, if there had been a third, fourth or even fifth backup available, he would have put those in place too, he replied, 'yes, I would have'. In this case, safety has become doing the best available under the circumstances. Again, this emphasises the overall level of trust in the system and their desire to prevent significant deviation from organizational norms. The operational managers assume that the overall risk management philosophy of the facility is appropriate and they aim to do the best that they can within that, rather than attempting to assess risks for themselves from first principles.

Story 7 is a similar case from the nuclear power station. In this case, prior to undertaking an unusual activity (blowing down the carbon dioxide system prior to maintenance), the operational manager realised that one of the normal safety barriers that would prevent high carbon dioxide levels in the reactor area was not present. This was the strong wind almost always blowing due to the exposed coastal location of the plant. Since he was unable to determine whether the wind was an important barrier in the safe design of the task (which had been done by others), he chose to put two extra temporary barriers in place (evacuating some parts of the site and positioning temporary hand-held detectors along with a system to stop venting if any increase in carbon dioxide was seen).

In a somewhat similar case at the air navigation service provider, one operational manager had to decide how to proceed when all the backup communications systems (backups for the system which allows air traffic controllers to speak to pilots) were put out of service by a hail storm (Story 22). In addition to instigating urgent repairs, he chose to bring in an extra staff member as a contingency in case the remaining communications system failed and to manage aviation traffic patterns around Sydney, to minimise the number of aircraft in the area with the communications system problem. The morning peak passed without further event with these extra safety barriers in place replacing the backup communications systems that were temporarily out of service. Similar to the chemical plant case, no comment can be made regarding the performance of these contingency arrangements since they were not called into play.

Other decisions at the air navigation service provider (summarised in Table 10.3) are also made considering barriers in place. In these cases, many of the issues with safety barriers are impacted by weather – either causing damage to systems, as in the case described above, or calling more safety systems into play, as in Story 20 and Story 21. External circumstances (in this case an aircraft damaging a runway) were also the cause of some temporary changes to safety barriers in Story 19.

Table 10.3   Air Traffic Control Story Summary – role of barriers

Story no.	Title	Role of safety barriers
18	Calibration and testing.	Manage maintenance so that barriers are reliable when they might be called upon.
19	Delaying arriving aircraft.	Use more reliable barrier (delayed departure rather than holding) despite schedule interruption.
20	Fix it now.	Failed barrier (instrument landing system) fixed quickly because it is needed.
21	Thunder storms approaching.	Improve operational barriers at a time when they are needed.
22	Hail in Canberra.	Failed barrier replaced with multiple temporary backups.
23	Traffic managers are a buffer.	Improve operational barriers at a time when they are needed.
24	Noise sharing.	Separation distance used as a barrier.
25	'When you kick a ball you don't know where it's going to land.'	Story shows that the effectiveness of barriers is a matter of judgement.
26	Third time unlucky.	Story demonstrates the importance of barriers (procedure for communications check).

It must be emphasised that no operational manager described his approach to safety in an abstract way in terms of ensuring barriers were in place (or risk controls, risk reduction measures or any similar generic term). Apart from declarative statements about the importance of safety, few interviewees attempted to describe in the abstract how they decided on a course of action: but the stories that they chose to illustrate what they consider safe had a common theme of barriers. This approach has been adopted to a significant extent by operational managers in all three organizations. For the purposes of operational decision-making, this approach allows managers to assume that, provided all safety barriers are in place and functional, then the system is safe. Whilst in a broad organizational sense a more critical view of existing safety systems is needed, for immediate operational issues in organizations with well-established systems this is a reasonable assumption to make. If some barriers are compromised, then operational managers adopt one of two options:

- stop/limit/curtail production to within the limits of the remaining barriers, or
- provide a temporary replacement barrier, which might be increased monitoring by the operational team.

The barriers approach is also an effective way of integrating compliance with rules with the operational managers' own expertise and judgement. In Story 2, for example, the barrier that was in danger of being breached was a rule – the maximum allowable concentration of a contaminant. The nuclear chemistry department sets this limit based on exposure limits. It is not an operational issue and the operational managers would be in no position to fix a maximum acceptable value based on their professional knowledge, as it falls under the purview of a different profession. Instead, once they are informed (by way of a procedure) that this limit is important, they proceed on that basis and use their operational expertise to manage the facilities within that limit.

Discussion of the results of risk assessment activities in terms of safety barriers (or controls) has become increasingly commonplace as part of a push to make the results of risk assessment more practical. This has led to the development of techniques such as bow tie analysis (Bice and Hayes 2009). In a small sample of offshore personnel, the UK Health and Safety Executive found that 'the workforce' liked the barriers approach to hazard management much more than 'management'. (HSE 2008) It is unclear who exactly is included in these two groups and hence which group would include the equivalent of operational managers, but the point appears to be that operations and maintenance field personnel identify more strongly with this approach than do office-based people with an engineering background. Part of the objection of the management group in the HSE study was that they felt the barriers approach was too difficult for the workforce to understand. This research suggests that, on the contrary at least at senior levels, the barriers approach is a generalisation of the way operations personnel think about managing hazards.

10.2 Risk

As discussed in Chapter 2, risk management is now synonymous with safety management for many organizations, and many organizations (including the three organizations that participated in this research) operate under a regulatory regime that requires them to reduce risk to a level that is as low as reasonably practicable (or ALARP). Thinking about devices and systems in place to control hazards as barriers has much in common with a risk management approach to safety, but also some key differences.

In technical terms, risk is a function of the potential consequences of an event, and the likelihood of those consequences occurring. Barriers can impact either or both of these parts of the risk equation in order to reduce or eliminate risk. In the risk management view of safety, each barrier is justified and prioritised based on its impact on risk. This is the basis of the justification that risk is as low as reasonably practicable and forms the nub of the discussion about critical risk controls (or barriers) required in some regulatory regimes. In this environment, safety decisions are often described in risk terms, for example, Reason links risk directly to safe decision-making when he defines a correct action (or decision) as 'one taken on the basis of an accurate risk appraisal' and an incorrect action as 'one based upon an inaccurate or inappropriate assessment of the associated risks'. (Reason 1997: 73) Discussions in research interviews with engineers who support operational decision-making were usually framed by interviewees in risk terms – discussion of frequency, consequence and acceptable level of risk. Once operational risk has been considered and an overall risk management strategy (based on a range of many barriers) has been put in place, then working with those barriers, rather than returning to a basic consideration of risk is a reasonable safety management strategy. The research data shows that operational managers assume that the system is safe if all barriers are in place as designed, and feel uneasy if barriers are compromised for any reason. One might think initially that this is consistent with a risk management view of decision-making because, after all, risk is a function of the barriers in place, but the subtle difference in emphasis can be significant.

The first implication of this focus on barriers rather than risk is that operational managers do not experience firsthand the trade-off between safety and cost that should have been considered when the barriers were selected in the first instance. Whilst the wider management of these conflicting goals rests with the organizations broadly (and ultimately the board of each organization), the original research design assumed that operational managers would be aware of the conflicting goals and would see their role as managers of this balancing act in the same way that resilience engineering theorists have focused on decisions balancing safety and production as one of the core competencies required of a resilient organization. Woods (2006) calls these decisions 'sacrifice judgements' – sacrificing production goals in the short term to ensure safety goals are met. In fact, operational managers at all sites saw their role quite differently. They saw their role largely in regard to safety as quite separate from their role in relation to production.

At the first research site (the chemical plant), some interviewees took exception to the suggestion in the research materials of any conflict between safety and production. At this site, the explanation given to participants at the beginning of each interview included this term. Several people thought they were being asked to recall occasions on which they had had an argument with their colleagues over a safety decision. Others understood the intent of the questions immediately but did not see that any conflict existed between safety and production. As interviewee 1 said, 'If it's a safety issue that requires us to shut the plant down, we do it. I don't see any conflict of interest at all. It's pretty cut and dried.' This appears to be a direct result of the operational managers' focus on the status of barriers, rather than a direct focus on assessing and evaluating risk.

The second implication of the focus on barriers is that operational managers may be slow to accept changes in organizational policies about safety barriers driven by reassessment of risks. There was one case of this seen at the air navigation service provider. At the time the research was carried out, operational managers were under pressure to reduce staff costs in one specific area. The operational requirement was such that commonly, for a period of around four hours in the late afternoon, the roster was short one controller. Under these circumstances, the typical arrangement in the past had been for a controller to be called in for an entire shift to cover this position, meaning that for half of that person's shift the roster would be overstaffed by one person. Management had instructed operational managers that this was not desirable and needed to be justified each time they made this decision. It was felt by senior management that the alternative operating arrangement (operating with one controller short and hence limiting the services provided to general aviation) was a better option – safe (by which they meant acceptable risk) and cheaper. The instruction on this issue was framed so that the final decision was always the responsibility of the operational manager, but the course of action preferred by management was clear.

The operational managers' responses to this varied. Some felt that they had been effectively instructed by more senior management not to bring in staff for extra hours and so they did not. Others felt that the decision was theirs to make and they were not happy to cut general aviation services (which include monitoring of search and rescue frequencies) in any circumstances simply to save on staff costs. They chose to call in staff anyway, even during daylight hours and perfect weather, when the need for search and rescue monitoring should be lowest. Traffic Manager 6 described this option as follows:'you put your head on the chopping block a lot and one day the axe will fall'.

An existing procedure detailed how the system should be operated (how to distribute tasks between available personnel and which services to suspend) with one fewer staff member than normal. The operational managers saw this as a contingency plan – a plan to manage an unforeseen short-term problem, rather than a description of a normal operating arrangement. On the other hand, the fact that the procedure described how to operate in this mode gave senior management confidence that this was a reasonable operational option.

Management decision-making in this case was based on their assessment of the trade-off between risk and cost – the risk to general aviation of operating for a short period without monitoring of the search and rescue radio frequency in one specific geographical area versus the cost of bringing in extra staff. The attitude of some operational managers was to consider only the loss of the safety barrier (the search and rescue monitoring). Whether or not, in hindsight, the search and rescue monitoring was actually called into action on any specific occasion was irrelevant to the operational managers, but of significant relevance to the more senior management position. It must be said that this situation was occurring during a period of industrial unrest, which was certainly colouring attitudes, if not actions, on all sides.

The barriers approach is practical and links directly to physical plant equipment and actions in the field. This may also explain why it is preferred over risk assessment as a decision-making frame. In an operational situation, risk assessment typically involves using a matrix (with consequence on one scale and likelihood on the other) in order to classify any given situation or option into one of three categories – intolerable, tolerable (or ALARP) and acceptable[1] (see Section 2.3.2). This is useful to operational managers only in extreme cases, when the risk is either intolerable or acceptable. Cases of clearly intolerable risk are usually obvious and would trigger immediate intervention from an experienced operational person. The problem is that most operational cases fall not at the extremes but in the middle category of the risk spectrum. For such cases, assessing risk as moderate or medium provides no guidance as to the best decision. At the end of the day, an operational manager must decide whether and how to intervene in the operating system from a range of medium risk options. In most cases the coarseness of the quantitative tool provides little or no guidance. The qualitative assessment of risk in an operational situation is often constructed in hindsight simply as a vehicle for expressing experience and judgement about a given situation in a more analytical form.

10.3 Summary and Concluding Comments

For operational managers, 'safe' means all safety barriers (or temporary alternatives) are in place. For organizations, 'safe' means that the risk associated with the situation is tolerable. These two concepts are similar but can differ in some important ways. Decisions related to barriers tend to be more clear-cut. If a barrier has failed or been breached, this will usually be clear to an experienced operational manager. The manager then has two alternatives: provide a backup/ alternative barrier or shut the system down. In the heat of dynamic operational situations, risk evaluations are much more nebulous and do not provide a clear path to the best way forward, except in the most extreme cases. A commitment to barriers is a more resilient approach (see Section 2.1).

When it comes to judging the results of specific decisions, the research data shows that, generally speaking, the operational managers and their organizational superiors base their judgement of the quality of operational decisions on the processes followed, rather than on outcome. Differences arise in the fact that the preferred process is different – barriers in place, rather than risk.

Operational managers rarely use risk-based concepts as a way of thinking about specific situations or deciding on actions – that is, they do not specifically consider chance, probability or likelihood of things going wrong. For them, safety is an active concept. Actions focus on two aspects:

- Compliance with rules, and
- Ensuring sufficient integrity of the barriers that prevent a specific hazard from becoming a reality (or keeping the beast in the box).

---

1 Or sometimes simply high, medium and low. Some organizations also have four categories of risk with "medium" essentially divided into two.

This provides some useful insights into considerations that should be taken into account in setting a line in the sand. Such questions as: What safety barriers are impacted by the current situation? Under what circumstances will it become unsafe to continue operation in this mode (For how long? Under what set of conditions?)

Chapter 11

Creating Environments for Better Decision-making

We have seen how operational managers working in a nuclear power station, a chemical plant and an air navigation service provider regularly make high stakes decisions that have the potential to impact on their own safety, the safety of their colleagues and/or the general public. They also have overall control and responsibility for short-term decisions about whether, and how, operations should continue. Operational managers work at the chaotic middle of the organization, where broad organizational exhortations, slogans and objectives such as 'safe, quality tonnes' and 'safety, orderly and expeditious' must be translated into dynamic management of moment-by-moment activities. We have seen that there is a range of organizational systems – both formal (such as rules and procedures) and informal (such as professional support from peers) – to assist operational managers in their decision-making and, at least in the organizations that took part in this research, they almost always get things right – apparently achieving the right balance between safety and production.

This work has identified factors that contribute to achieving good outcomes in these complex and dynamic workplace environments. The key to understanding decision-making practices is sensemaking, a perspective which highlights the role of identity construction in decision-making – the differing contributions and potential tensions between the identity of the operational managers as employees and their identity as professionals.

Friedson (2001) calls professionalism the third logic. He contrasts a professional way of organizing work with two other modes of organizing work that he claims are now more common in our society. The first is an ideal type based on the free market, where customers and market demand drive prices, and hence priorities, in organizations that aim to maximize profits. The second is a bureaucratic form of organizing where reliability and predictability are organizational goals. Adherence to rules and monitoring of activities by organizational superiors dominate.

Whilst any real workplace or organization is likely to include elements of all three ideal types (professionalism, free market and bureaucracy), Friedson claims that professionalism is undervalued, and findings here support this. His general perspective can be applied to the balance between safety and production in high-hazard organizations. Common organizational views of how this balance is achieved are based around risk (which is fundamentally a market-driven approach trading off cost against safety) and rules (a bureaucratic approach), but the research has shown that professionalism plays an important and undervalued role in operational decision-making. Organizations implicitly acknowledge the professional status of operational managers by giving them operational authority that sits outside the organizational hierarchy, yet the broader implications of this sense of professionalism are rarely acknowledged directly or discussed.

The implications for safety decision-making are discussed below, followed by some suggestions for how organizations may support their operational managers in making better decisions.

11.1 Conclusions

Each of the organizations studied has in place a written system for management of work based on the general principles laid down in quality management standards, especially those related to the value of defining processes to achieve consistent outcomes. Rules in the form of procedures are seen by management as an effective and appropriate way to control the activities of their employees in order to achieve desired outcomes, including excellent safety performance. This applies broadly across each organization, but also specifically to operational activities. HRO theory has highlighted the limitations of this approach in situations where learning by trial and error is not an appropriate strategy, but the quality approach remains a key management philosophy in many industrial organizations, including the three that took part in this research. Rule-based approaches of various kinds are also promoted both by industry standards and by regulation in all three industries. In many cases, this approach works well. There are clearly occasions in each organization where operational managers used pre-existing rules and procedures when making safety-related decisions. In most cases, these are rules relating to system operating limits (rather than rules specifying a process to be followed or a specific concrete action or state to be achieved).

In each organization, some rules are in place specifying limits related to maximum or minimum values of various individual physical parameters, such as maximum operating pressure, minimum operating temperature, minimum allowable separation distance. In the case of the chemical plant and the nuclear power station, formal limits also include specifications for the minimum equipment with a key safety function that must be on line or available at any given time. In all three industries, published standards promote the setting of such fixed limits as a key safety strategy. These limits are based on formal analysis and are often prescribed by technical/engineering disciplines other than those directly involved in system operation. This is a key way in which limits on operations inherent in, for example, the design of the system can be translated into the field.

Such pre-determined safety limits provide important information to operational managers about the safety of any given abnormal situation. In practice, however, whilst exceeding a specified limit was known to be unsafe, operating *within* the fixed limits was not seen by operational managers as a universally safe option. There were many examples given in interview where managers relied on experience-based judgement, rather than recollection of a published rule, in deciding that a significant production interruption was necessary. Such cases included:

- Deviation from the expected value of more than one parameter, where each individual parameter remained within its stipulated operating envelope;
- Response to an unplanned equipment failure or outage; and
- Situation-specific knock-on effects of compliance with a safety rule.

In some of these cases, operational managers created a new, situation-specific, goal-based rule for themselves in the form of a 'line in the sand'. These self-imposed limits were often a maximum acceptable time for operation in an abnormal state, but could sometimes be another system parameter. The nuclear power station had the most developed system for this, where operational managers set a line in the sand in consultation with the operating team, documented it and then worked in accordance with that new self-imposed limit. In these cases, the sense that was made of the developing problem was used to create a new structured environment for further sensemaking which appears to prevent normalization of the abnormal situation and the potential for incremental movement away from normal operating practices to less and less conservative options.

The research has shown that senior managers (higher in the organizational hierarchy than the operational managers) often have a relatively poor understanding of the factors driving decision-making by operational managers, and that this second key aspect to decision-making enacted within specified safety limits was largely invisible to the broader organization. The reliance on rules, whilst a significant factor in decision-making at all sites, is not as all-encompassing as many senior managers believe. They appear to have been seduced to a significant extent by the charm of written rules and procedures. The remoteness of senior management from immediate operating issues allows them to experience 'the sense of certainty available in theory, so welcome compared to the unsettled uncertainty and anxiety of decision that pervades the realm of practice'. (Sullivan 2005: 245) On the other hand, operational managers well understood the uncertainty and unpredictability of the systems within which they operate. As one said, 'when you kick a ball, you don't know where it's going to land' that is, the outcome of a given course of action cannot be uniquely determined in advance, even by an expert. The complexity and variability in the system are such that all any individual can do is to work to improve any given situation and to continue to monitor developments in case further actions are needed.

In these types of situations, the operational managers turn to their experience and judgement to resolve the best way forward. In this mode of thinking, stories, rather than analysis, prevail. All operational managers were adamant that 'if it's not safe, we don't do it'. When asked to describe or generalize about how they decided whether a given situation was safe or unsafe, interviewees rarely talked in theoretical or generalized terms, but often recounted stories about their past experiences – in particular, occasions that had brought home to them the reality of the danger associated with the technologies with which they work. These stories covered a range of incidents and accidents, large and small. The common feature was that each event had challenged the storyteller's understanding of how the system operates and the degree of control they have over it. This sense of chronic unease and the need for constant vigilance (Reason 1997: 37) was also reflected in the language used to describe the system. Several managers characterized the system that they work with as a beast – something that is dangerous, unpredictable and needs to be controlled.

When considering their own course of action in the face of a developing operational issue, some operational managers described using story-based tests. These tests take a number of different forms. In one type of test, the decision-maker asks himself: would I want my family to be here? In other cases, he imagines himself to be at some point in the future and asks himself: would I be happy explaining this to the family of someone who was injured, or would I be happy for my part in this to be published in the newspaper? No matter which type of test is used, the conclusions are based on the storyteller's emotional response to the situation he imagines himself to be in. This form of creative, rather than analytical, thinking provides the alternative in these cases to trial and error in the field and is therefore an important facet of moving to appropriate action in the high hazard environment.

Stories are also used to make sense of, and share, technical knowledge about the system. All interviewees are very experienced individuals. Many have spent their professional career within one organization. Almost all have over 20 years' experience within their industry. As would be expected of such a group, their continued learning is based on highly specific experiences and stories that increase or reinforce their understanding of how their particular system behaves. The operational managers are keenly interested such incidents that add to their factual understanding but also, critically, increase their ability to imagine how the system is likely to behave in situations that they may not have experienced. Sharing stories of recovery from unusual operating modes and difficult situations also fosters the sense that, despite our best attempts, not all unusual system behaviours can be identified in advance and hence planned for. Operating excellence also requires operational managers to have the ability to respond and adapt to circumstances in order to recover control and stories help them achieve this. These are the third type of stories that contribute to decision-making by operational managers.

Each of the organizations we have examined has a comprehensive system in place for reporting, investigating, acting upon and trending incidents and accidents. Whilst not the specific focus of this research, the systems appear to play an important role in resolving specific safety issues in a timely manner and in identifying longer term trends and opportunities for improvement generally. Reporting rates at all sites are high. Whilst the operational managers use all aspects of these systems on a daily basis, the incidents of interest to them in increasing their system understanding are usually not reported or reportable in the site or organization-wide system, because the focus of reporting systems is on incidents related to loss or potential loss (in the case of the chemical plant and nuclear power station) and error (in the case of the air navigation service provider). On the other hand, incidents relating to unusual and unexpected system interactions and events are of interest to operational managers. Whilst some incidents fall into both categories, many do not. In each organization, operational managers had developed informal systems outside the main incident reporting arrangements for sharing their experiences in the form of detailed stories.

We have seen three ways in which operational managers use and share stories as part of their professional identity:

* As a way of remembering what an unsafe situation feels like,
* As a way of imagining the future and hence testing a possible course of action, and
* As a way of sharing 'in practice' technical knowledge with their peers.

As professionals, they act from a sense of integrity and public trust, taking their authority from their expertise, experience and knowledge. Their professionalism gives them a sense of independence from the non-specialist requirements of their employer, and a sense of dedication to their profession that they share with their professional peers. It has several other important implications for safety decision-making.

The first relates to the issue of production pressure. Operational managers in all three organizations were well aware of the financial environment within which their organizations operate and the importance of meeting production targets and keeping costs down. In two of the three organizations, operational managers felt very well supported by senior management in making 'sacrifice decisions' where these longer term goals must sometimes be traded off against short term safety concerns. As employees, they had no problem with making that judgement. The pressure to stay online in the face of developing operational difficulties seems to come from their own professional pride. Operational managers can sometimes put themselves under pressure as a result of the way they see their primary task – producing electricity, making plastic or moving aircraft. They are reluctant to be thwarted by the system in performing that task and can see a need to shut down as a professional failure. Returning to the beast metaphor described above, the operational managers see their role as being to contain and control the beast, but not to kill it.

It is within this context that support from more senior managers is very important. Simplistically, a good decision from an operational manager might be seen as disaster averted, but this introduces some major problems if used to determine the effectiveness of real decisions. There are some cases in the research data where a decision was made to interrupt production and further analysis showed that there was indeed a significant fault in the system that might have resulted in a serious incident if no intervention had been made. In these cases, it is clear that a real problem was averted, but many cases do not conform to this pattern and yet the decisions can still be classified as good ones. In some cases, the fact that the intervention occurred makes it impossible to know what would have happened otherwise (if the aircraft had not been grounded as the bad weather developed at their destination, what would have happened?). Whether any individual judgement was correct in context or overly conservative remains a matter of opinion, even after the event. In other cases, further analysis shows that, however appropriate the judgement to intervene seemed at the time, there was in fact no serious threat to safety and the interruption to production was unnecessary when judged in hindsight by the actual outcome. In two of the three organizations, operational managers felt that their decision would always be supported by their manager, even in the case of these types of decisions.

Also in two of the three organizations studied, roles related to budget accountability had been deliberately defined so as to separate responsibility for safety decisions from responsibility for the associated costs. An example of this is that operational managers in each organization have the authority to call in a maintenance crew outside of normal hours due to safety (or operational) concerns, but the overtime costs are managed elsewhere within the organization so that the decision-maker is not literally faced with the bill.

Whether pressures arise from professional pride, or from a need to meet organizational goals, the research shows that overall success from the perspective of both professionals and employees must come from putting short term safety problems above production and cost imperatives.

Whilst the operational managers could not provide a general explanation of the reasoning behind their safety decisions, a review of the research data gives some insights into how they appear to make this judgement. In the specific decision-making scenarios described in interview, individuals focused on context-specific loss of barriers (or risk controls, or defences) that is, deviation from the normal level of control of the hazard. If the loss of barriers was not so severe as to lead to an immediate decision to shut down, then actions were typically three-fold:

- Firstly, define a 'line in the sand' – usually repair time or a limit on a parameter that defines further degradation of a safety barrier (such as falling pressure in a seal system),
- Secondly, put temporary or alternative systems in place as a backup to the failed barrier, and
- Thirdly, monitor the situation and make any changes to these strategies, as new evidence requires.

The role of their experience and judgement is in noticing the degraded or failed barrier in the first instance, in understanding the significance of the failure, in finding short term alternatives and in fixing the line in the sand. This barriers approach is not inconsistent with broader organizational risk management (which should have been the way the design of the various barriers was developed in the first place), but explains why operational managers do not use risk directly in reaching safety decisions. Consideration of barriers is less subjective than trying to assess risk in a dynamic operational environment and explains why operational managers often do not see a technical conflict between safety and production. If safety is a matter of barriers in place then no direct trade-off with production is necessary.

In many ways it must be obvious that the professionalism of the operational managers as a group, encompassing such aspects as their judgement, skills and experience and commitment to the job, plays an important part in good decision-making. This raises the question as to whether it matters that these factors are not openly acknowledged. The fact that operational managers act as professionals as well as from their identity as employees *should* matter to organizations seeking excellent safety performance. If judgement-based processes are acknowledged and valued, then those judgements can be refined and improved – to the benefit of all.

11.2 Implications for Operating Companies

This work has not resulted in the development of a five (or even ten) factor model of good decision-making. Instead, the aim is to provide risk case study material in order to foster readers' understanding of the complexity of operational decision-making and the richness of existing informal ways in which this knowledge is shared. Any review of the causes of accidents in complex systems shows that organizations ignore this issue at their peril. The following sections describe a number of practical ways in which organizations may assist operational managers to make better safety decisions.

11.2.1 Clarifying the boundary of the system operating envelope

Consistent decision-making is supported by a shared understanding of what constitutes 'unsafe' in the sense of the operating envelope of the system. The range of parameters that need to have limits assigned will vary from facility to facility, but will typically include:

- Physical parameters such as pressure, temperature, composition in the case of the process plants and traffic levels in the case of the air navigation service provider,
- Other internal parameters such as minimum numbers of people with certain skills required to cover the complete range of possible operational conditions, including emergencies,
- External environmental parameters such as maximum or minimum weather conditions,
- Limits on availability of safety systems or other equipment items, and,
- Limits on acceptable combinations of simultaneous activities.

Whilst operational managers and others on the operating crew have important insights into the operating limits of the system, some limits are inherent in the design. These are generally engineering issues that may not be at all obvious without reference to the original design calculations. Such calculations are explicitly, or implicitly, based on risk assessment. This information must be communicated to the operational crew, and fixing operational boundaries is one simple way to achieve this.

As the research has shown, whilst operation outside the design envelope may be unsafe, operational managers also identify many cases where they believe that operations not specifically proscribed are still potentially unsafe. Such cases are not an indication that the operating envelope has been incorrectly defined, but simply a sign that not every operating mode of a dynamic and complex system can be predicted in advance. For rules regarding operating limits to be accepted by operational managers, it is important to emphasize that they are designed to support operational decision-making rather than completely define it and that judgement remains an important aspect in decision-making.

11.2.2 Formalizing judgement-based decision processes – the line in the sand

The logical way to acknowledge the place of judgement and experience in operational decision-making is in the form of process rules for decision-making. The decision-making process defined at the chemical plant was not accepted by the operational managers, because it makes no allowance for the role of experience and judgement. The senior management focus on classical decision-making processes did not sit well with the cognitive processes already in use. On the other hand, the concept of conservative decision-making as defined at the nuclear power station was universally accepted by operational managers precisely because it provided a formal framework for their experience. The steps in this process once an unusual operating condition has been detected are:

- Assess the current situation and decide if an immediate production interruption is required,
- If not, commence troubleshooting or repair, and also set a 'line in the sand' that defines the point at which troubleshooting will end and production will be interrupted,
- Record and communicate the 'line in the sand' to all involved,
- Monitor the system for further changes to ensure that the 'line in the sand' is still appropriate, and,
- If the line in the sand is reached before the situation returns to normal, then initiate whatever production interruption has been previously agreed upon.

It is telling that operational managers at the chemical plant and at the air navigation service provider had adopted at least the first two steps of this process for themselves, including coming to a conclusion about the time they had available to fix a problem. The steps in the process that were not always applied were recording, communicating and sticking to the self-imposed limit that the operational manager had determined when the operating anomaly first came to his attention. Instead, they had a tendency to allow the abnormal situation to continue, especially if there was no evidence that the situation was deteriorating. This provides strong evidence that what Vaughan (1996) calls 'the normalization of deviance' occurs readily in operating situations and that a formal process to stick to self-imposed limits could limit the drift towards acceptance of unsafe conditions.

The research suggests that operational safety would be enhanced if operational managers were trained to set a line in the sand and then stick to it as part of their troubleshooting in unusual operating situations. They should not be expected to continue operating simply because the system is within the defined operating envelope. This decision-making process seems to provide a structure and support for professional judgement, rather than defining cognitive steps that do not come naturally to experienced decision-makers under time pressure.

11.2.3 Sharing professional knowledge

As described in Section 8.3, operational managers love to share with their peers stories of their experiences. This currently happens in all three organizations studied – but outside formal incident reporting systems and also with no specific time allocation. As a group, the behaviour of the operational managers at each site reflects many of the key qualities of Wenger's *community of practice* (1998), a concept that has been widely adopted in the field of knowledge management, and there are opportunities for organizational structures, priorities and systems to support this.

The scope of incident reporting systems is usually limited to loss or potential loss events. The types of stories of interest to operational managers about interesting system behaviours often do not fall into this category. They are stories about unusual operating occurrences that may have only an indirect link to loss or loss prevention. None of the organizations studied systematically records such incidents so that lessons can be captured for the longer term and incorporated into formal training and/or simulation. Such stories support development of a deep understanding of system behaviours and hence have an important role in long term safe operation. Operating personnel generally have less hands-on experience of the system than in the past due to the high degree of automation for complex plant and systems. This makes the ability to anticipate problems and know how the system will behave is becoming more and more difficult to maintain.

In modern organizations, all staff are under time pressure and time for unacknowledged tasks is becoming more and more difficult to find. It was common years ago to have regularly scheduled training days, when operational staff could get together away from day-to-day pressures to discuss items of common professional interest. Such sessions have tended to suffer from cost reduction pressures or to be taken over by organizations as time to promote the latest management programs, rather than professional development time for operational staff. High reliability theory describes the benefits of both seeking diverse views and valuing deep and idiosyncratic operational understanding. It is possible that electronic networking tools could provide an alternative to face-to-face meetings and play an important role in sharing of stories, but it must be acknowledged that professional qualities take time to foster and develop. In a commercial environment where time is seen as money, recognizing the role of professional judgement in decision-making might make it easier to justify the costs associated with professional development time and prevent such time from being taken over by other organizational communications requirements.

11.2.4 Consultation

In a similar vein, operational managers at all three organizations were very happy to assist their colleagues by receiving out of hours phone calls and requests for assistance. This was largely unacknowledged by the wider organization. All three organizations also have emergency callout arrangements that are designed to be triggered in the event of a serious emergency at the site. People who are included on the emergency roster have allocated emergency roles and have been trained to fill them. At the nuclear power station, this system had been adapted gradually and informally to include consultation in non-emergency situations. This has created some anomalies regarding training for inclusion on the roster and remuneration for duties.

If consultation is to be encouraged, then it should be resourced and appropriate rewards provided. This is not to say that specifics of consultation between professional colleagues need to be authorized or controlled by senior management. It is simply that modern business practices, such as business process definition and re-engineering, give no value to activities that are not identified and seen as being in accordance with organizational business needs and, in this case, tools and time need to be allocated.

11.2.5 Clarifying goals and how to achieve them

Operational managers in all three organizations were well aware of the financial environment within which their organizations operate, in particular the importance of meeting production targets and keeping costs down. On the other hand, they take their safety responsibilities very seriously and are prepared to make 'sacrifice decisions' where these longer-term goals must sometimes be traded off against short-term safety concerns.

One of the most strongly articulated issues about safety decision-making was the attitude of senior management, specifically the response to particular decisions made. It is important for senior managers to foster a common understanding with operational managers of overall safety and production goals and then to develop and demonstrate their trust in the ability of the operational managers to achieve those goals. It can be too easy for senior management to espouse the overall importance of safety being 'our number one priority', without acknowledging what this means on a day-to-day basis. If safety is truly to be top priority, then complexity and uncertainty will mean that there are sometimes legitimate production interruptions in cases where hindsight will show the interruption was unnecessary.

Operational managers in all three organizations were very sensitive to senior management responses in these situations.

11.2.6 Emphasis on barriers, rather than risk

Risk assessment (consideration of hazards, consequences and frequency) plays an important role in establishing the safe operating envelope of the system and in determining what risk controls or barriers should be included in the system as a whole. On the other hand, the concept of risk appears to have little impact on day-to-day operational decision-making. This is not a recommendation for more training or additional focus on risk assessment techniques for operational staff, but rather an observation about the limits of the concept of risk in practical situations.

Formal risk management processes are ultimately aimed at ensuring that sufficient controls are in place to ensure that risk is as low as reasonably practicable (ALARP). With this in mind, many operational risk management regimes are moving their focus from risk per se to risk controls or barriers. This is more consistent with the way in which operational personnel think about safety and perhaps explains the popularity of risk management tools such as bow ties (which focus on barriers rather than on estimates of risk) with operational personnel. If overall risk management processes focus on ensuring that the right risk barriers are in place for normal operations, then the decision-making strategy of looking at deviations from that set of barriers seen in this research data will be a successful risk management approach. This leads directly to a procedure for operational decision-making based on the line in the sand approach detailed in Section 11.2.2, rather than use of operational risk matrices or similar.

11.3 Implications for Regulators

In addition to providing pointers for operating organizations, these case studies are also relevant to regulators working in high-hazard sectors. Regulators tend to focus on documentation and records in their efforts to judge compliance with regulations that typically focus attention on written safety management systems. Experience across a range of industries and our three case studies suggests that regulatory involvement with system operations is low unless and until there is an incident of some kind. Much of what happens in day-to-day operations is not documented in any form that is auditable, particularly by non-specialists, so it is easier for regulators to focus on other areas such as incident reporting, inspection programs and maintenance records. These areas have been shown to be important and the document trail lends itself to an audit-based judgement by regulators regarding compliance.

This research provides the basis for regulatory enquiry into operations and provides guidance for regulators as to what they should look for in answer to questions such as:

- What are the operational limits of your system that are inherent in the design of the system and how is this information communicated to your operations personnel?
- What actions are required by operations personnel driven by these limits?
- How do operations personnel make safety decisions that are not driven directly by these limits?
- To what extent does your organization acknowledge its reliance on the professional judgement and experience of your operations personnel?
- How does your organization support the professionalism of your operations personnel?
- How are unusual operating experiences recorded and shared?
- How are the results of system-wide base case risk assessments linked to day-to-day decision making?

In closing, we have seen that the operational managers in the nuclear power station, the chemical plant and the air navigation service provider are stable long-term employees who show clear respect for organizational goals in making operational decisions. They also have a rich professional life that is largely ignored by formal organizational processes. Their organizational experience provides insights into the importance of their identity as professionals, as well as employees, in safety decision-making.

Operational managers have described making safety decisions by drawing a line in the sand – that is, creating a short-term situation-specific rule in conjunction with their colleagues and then using this rule to drive action. This is judgement-based, time-pressured, dynamic decision-making and yet, at its best, it is still disciplined. Details are recorded for later consideration and sharing in the form of a vivid story of the events that occurred and the choices that were made. This work has highlighted ways in which all organizations operating in a high-hazard environment could support their own operational managers in making better safety decisions by acknowledging and supporting their professional experience and judgement.

References

Abt, E., Rodricks, J.V., Levy, J.I., Zeise, L. and Burke, T.A. 2010. Science and Decisions: Advancing Risk Assessment. *Risk Analysis,* 30(7), 1028–1036.

Airservices Australia. 2006. Media Release, Fatal Accident involving VH-TNP near Benalla on 28 July 2004. [Online] Available at: http://www. airservicesaustralia.com/media/press_releases/pr.asp?id=pr1_06 [accessed: 18 August 2008].

Airservices Australia. 2007. Annual Report 2006–2007. [Online] Available at: http://www.airservicesaustralia.com/wp-content/uploads/Airservices_Annual _Report_2006-2007.pdf [accessed: 30 April 2012].

Argyris, C. 2004. *Reasons and Rationalisations: The Limits to Organizational Knowledge,* Oxford: Oxford University Press.

Australian Transport Safety Bureau. 2006. Piper Aircraft Corp PA-31T, VH-TNP, Aviation Occurrence Report – 200402797. [Online] Available at: http://www. atsb.gov.au/publications/investigation_reports/2004/aair/aair200402797.aspx [accessed: 18 August 2008].

Australian Transport Safety Bureau. 2007a. Australian Aviation Safety in Review: 2002–2006. Aviation Research – AR-2007-061 Second Edition.

Australian Transport Safety Bureau. 2007b. Crosswind Landing Event, Melbourne Airport, Vic, 26 October, 2005, HS-TNA, Airbus A340-642, Aviation Occurrence Report – 200505311. [Online] Available at: http://www. atsb.gov.au/publications/investigation_reports/2005/aair/aair200505311.aspx [accessed: 30 April, 2012].

Australian Transport Safety Bureau. 2007c. Trends in Immediately Reportable Matters Involving Regular Public Transport Operations (Aviation Research and Analysis Report - B20070107). Canberra: Commonwealth of Australia.

Crowded Skies: The Blame Game, 2003. Directed by BBC. UK.

Beamish, T. 2002. *Silent Spill, The Organization of an Industrial Crisis,* Cambridge, MA: The MIT Press.

BFU 2004. Investigation Report AX001-1-2/02. Braunschweig: German Federal Bureau of Aircraft Accident Investigation.

Bice, M. and Hayes, J. 2009. Risk Management: From Hazard Logs to Bow Ties, in *Learning from High Reliability Organisations*, edited by A. Hopkins. Sydney: CCH.

Bluff, L. and Johnstone, R. 2004. The Relationship Between 'Reasonably Practicable' and Risk Management Regulation. *National Research Centre for OHS Regulation.* Canberra.

Bourrier, M. 1996. Organising Maintenance Work At Two American Nuclear Power Plants. *Journal of Contingencies and Crisis Management,* 4(2), 104–112.

Bourrier, M. 2002. Bridging Research and Practice: The Challenge of 'Normal Operations' Studies. *Journal of Contingencies and Crisis Management,* 10(4), 173–180.

Bourrier, M. 2005. An Interview with Karlene Roberts. *European Management Journal,* 23(1), 93–97.

Bourrier, M. 2011. The Legacy of the High Reliability Organization Project. *Journal of Contingencies and Crisis Management,* 19(1).

BP. 2010. Deepwater Horizon Accident Investigation Report September 8 2010. [Online] Available at: http://www.bp.com/sectiongenericarticle.do?categoryId =9040069&contentId=7067574 [accessed: 12 June 2012].

Carroll, J.S. 1993. Out of the Lab and Into the Field: Decision Making in Organizations, in *Social Psychology in Organizations,* edited by J.K. Murninghan. Englewood Cliffs, NJ: Prentice Hall, 38–62.

Carroll, J.S. 1998. Organizational Learning Activities in High-hazard Industries: The Logics Underlying Self-analysis. *Journal of Management Studies,* 35(6), 699–717.

Carroll, J.S., Rudolph, J. and Hatakenaka, S. 2002. Learning from Experience in High-Hazard Organizations, in *Research in Organizational Behavior,* edited by B. Staw & R. Kramer. Oxford: Elsevier Science.

Carroll, J.S., Rudolph, J., Hatakenaka, S., Wiederhold, T. and Boldrini, M. 2001. Learning in the Context of Incident Investigation: Team Diagnoses and Organizational Decisions at Four Nuclear Power Plants, in *Linking Expertise and Naturalistic Decision Making,* edited by E. Salas & G. Klein. Mahwah, NJ: Lawrence Erbaum Associates.

Carroll, J.S., Sterman, J. and Marcus, A. 1998. Playing the Maintenance Game: How Mental Models Drive Organizational Decisions, in *Debating Rationality: Nonrational Aspects of Organizational Decision Making,* edited by J.J. Halpern & R. Stern. Ithaca: Cornell University Press.

Causer, G. and Jones, C. 1996. One of Them or One of us? The Ambiguities of the Professional as Manager, in *New Relationships in the Organised Professions,* edited by R. Fincham. Aldershot: Ashgate.

Center for Chemical Process Safety 2007. *Guidelines for Risk-Based Process Safety,* Hoboken, NJ: Wiley Interscience.

Chandler, A.D. 1977. *The Visible Hand: The Managerial Revolution in American Business,* Cambridge, Mass.: Harvard University Press.

Cohen, M.D., March, J.G. and Olsen, J.P. 1972. A Garbage Can Model of Organizational Choice. *Administrative Science Quarterly,* 17(1), 1–25.

Cullen, L. 1990. *The Public Inquiry into the Piper Alpha Disaster.* London: HMSO.

Dawson, D.M. and Brooks, R.J. 1999. *The Esso Longford Gas Plant Accident: Report of the Longford Royal Commission.* Melbourne: Parliament of Victoria.

Defoe, J. and Juran, J.M. 2010. *Juran's Quality Handbook: The Complete Guide to Performance Excellence*, New York: McGraw Hill.

Dekker, S. 2004. Why We Need New Accident Models. *Human Factors and Aerospace Safety*, 4(1), 1–18.

Deming, W.E. 2000. *Out of the Crisis*, Cambridge: The MIT Press.

Dreyfus, H.L. and Dreyfus, S.E. 1986. *Mind over Machine*, New York: The Free Press.

Etzioni, A. 1967. Mixed-Scanning: A 'Third' Approach to Decision Making. *Public Administration Review*, 27(5), 385–392.

Flin, R. 1996. *Sitting in the Hot Seat, Leaders and Teams for Critical Incident Management*, Chichester: John Wiley & Sons.

Flyvbjerg, B. 2001. *Making Social Science Matter: Why Social Inquiry Fails and How it Can Succeed Again*, Cambridge: Cambridge University Press.

Friedson, E. 2001. *Professionalism: The Third Logic*, Chicago: University of Chicago Press.

Hale, A. and Heijer, T. 2006. Defining Resilience, in *Resilience Engineering: Concepts and Precepts*, edited by E. Hollnagel, D.D. Woods & N. Leveson. Aldershot: Ashgate.

Hale, A.R. and Swuste, P. 1998. Safety Rules: Procedural Freedom or Action Constraint? *Safety Science*, 29, 163–177.

Hayes, J. 2009. Incident Reporting: A Nuclear Industry Case Study, in *Learning from High Reliability Organisations*, edited by A. Hopkins. Sydney: CCH.

Hayes, J. 2012. Operator Competence and Capacity – Lessons from the Montara Blowout. *Safety Science*, 505, 63–574.

Hollingworth, A. 2007. Dealing with Stress at Work. *Sunday Life*.

Hollnagel, E. 2006. Resilience – the Challenge of the Unstable, in *Resilience Engineering: Concepts and Precepts*, edited by E. Hollnagel, D.D. Woods & N. Leveson. Aldershot: Ashgate.

Hollnagel, E. 2008. Preface: Resilience Engineering in a Nutshell, in *Remaining Sensitive to the Possibility of Failure*, edited by E. Hollnagel, C.P. Nemeth & S. Dekker. Aldershot: Ashgate.

Hollnagel, E. 2009. *The ETTO Principle: Efficiency-Thoroughness Trade-off, Why Things That Go Right Sometimes Go Wrong*, Farnham: Ashgate.

Hollnagel, E., Nemeth, C.P. and Dekker, S. (eds.) 2008. *Remaining Sensitive to the Possibility of Failure*, Aldershot: Ashgate.

Hollnagel, E., Pariès, J., Woods, D.D. and Wreathall, J. (eds.) 2011. *Resilience Engineering in Practice: A Guidebook*, Farnham: Ashgate.

Hollnagel, E., Woods, D.D. and Leveson, N. (eds.) 2006. *Resilience Engineering: Concepts and Precepts*, Aldershot: Ashgate.

Hopkins, A. 1999. The Limits of Normal Accident Theory. *Safety Science*, 3293–102.

Hopkins, A. 2000. *Lessons from Longford: The Esso Gas Plant Explosion*, Sydney: CCH.

Hopkins, A. 2001. Was Three Mile Island a 'Normal Accident'? *Journal of Contingencies and Crisis Management*, 9(2), 65–72.

Hopkins, A. 2005a. The Gretley Coal Mine Disaster: Reflections on the Finding that Mine Managers were to Blame. *National Research Centre for OHS Regulation.* Canberra.

Hopkins, A. 2005b. *Safety, Culture and Risk: The Organisational Causes of Disasters,* Sydney: CCH.

Hopkins, A. 2006. Studying Organisational Cultures and their Effects on Safety. *In:* International Conference on Occupational Risk Prevention, 2006 Seville.

Hopkins, A. 2008. *Failure to Learn: The BP Texas City Refinery Disaster,* Sydney: CCH.

Hopkins, A. 2009a. Identifying and Responding to Warnings, in *Learning from High Reliability Organisations,* edited by A. Hopkins. Sydney: CCH.

Hopkins, A. (ed.) 2009b. *Learning from High Reliability Organisations,* Sydney: CCH.

Hopkins, A. 2012. *Disastrous Decisions: The Human and Organisational Causes of the the Gulf of Mexico Blowout,* Sydney: CCH.

HSE 2001. Reducing risks, protecting people: HSE's decision-making process. Norwich: UK Health and Safety Executive.

HSE 2008. Optimising hazard management by workforce engagement and supervision. Norwich: UK Health and Safety Executive, Research Report RR637.

Hudson, P.T.W. 1999. Safety Culture – Theory and Practice. *In:* Human Factors and Medicine Panel (HFM) Workshop – The Human Factor in System Reliability: Is Human Performance Predictable? 1999 Siena Italy. North Atlantic Treaty Organization Research and Technology Organization.

ILO 2001. Guidelines on occupational safety and health management systems ILO-OSH 2001. Geneva: International Labour Office.

International Atomic Energy Agency 2000. *Operational Limits and Conditions and Operating Procedures for Nuclear Power Plants. Safety Standards Series No. NS-G-2.2,* Vienna: IAEA.

Janis, I. 1982. *Groupthink: Psychological Studies of Policy Decisions and Fiascos,* Boston: Houghton Mifflin.

Kepner, C.H. and Tregoe, B.B. 1997. *The New Rational Manager,* Princeton: Princeton Research Press.

Kjellen, U. 2000. *Prevention of Accidents Through Experience Feedback,* London: Taylor & Francis.

Klein, G. 1998. *Sources of Power: How People Make Decisions,* Cambridge, Massachusetts: MIT Press.

Klein, G. 2003. *The Power of Intuition: How to Use Your Gut Feelings to Make Better Decisions at Work,* New York: Currency/Doubleday.

Klein, G.A. 2009. *Streetlights and Shadows: Searching for the Keys to Adaptive Decision Making,* Cambridge: MIT Press.

La Porte, T.R. 1981. On the Design and Management of Nearly Error-Free Organizational Control Systems, in *Accident at Three Mile Island: The Human*

Dimension, edited by D.L. Sills, W.C.P. & V.B. Shelanki. Boulder, CA: Westview Press.

La Porte, T.R. 1996. High Reliability Organizations: Unlikely, Demanding and At Risk. *Journal of Contingencies and Crisis Management,* 4(2), 60–71.

La Porte, T.R. and Consolini, P.M. 1991. Working in Practice but Not in Theory: Theoretical Challenges of 'High-Reliability Organizations'. *Journal of Public Administration Research and Theory,* 1(1), 19–48.

Lakoff, G. and Johnson, M. 1980. *Metaphors We Live By,* Chicago: University of Chicago Press.

Laroche, H. 1995. From Decision to Action in Organizations: Decision-Making As a Social Representation. *Organization Science,* 6(1), 62–75.

Lipshitz, R., Klein, G., Orasanu, J. and Salas, E. 2001a. Focus Article: Taking Stock of Naturalistic Decision Making. *Journal of Behavioural Decision Making,* 14331–352.

Lipshitz, R., Klein, G., Orasanu, J. and Salas, E. 2001b. Rejoinder: A Welcome Dialogue – and the Need to Continue. *Journal of Behavioural Decision Making,* 14385–389.

McAdams, D.P. 1993. *The Stories We Live By: Personal Myths and the Making of the Self,* New York: The Guilford Press.

Middlehurst, R. and Kennie, T. 1997. Leading Professionals: Towards New Concepts of Professionalism, in *The End of the Professions? The Restructuring of Professional Work,* edited by J. Broadbent, M. Dietrich & J. Roberts. London: Routledge.

Morgan, G. 1997. *Images of Organization,* Thousand Oaks: Sage Publications.

Ocasio, W. 2005. The Opacity of Risk: Language and the Culture of Safety in NASA's Space Shuttle Program, in *Organization at the Limit: Lessons from the Columbia Disaster,* edited by W. H. Starbuck & M. Farjoun. Malden, MA: Blackwell.

Pariès, J. 2011. Resilience and the Ability to Respond, in *Resilience Engineering in Practice: A Guidebook,* edited by E. Hollnagel, J. Pariès, D., Woods, D. & J. Wreathall. Aldershot: Ashgate.

Perin, C. 2005. *Shouldering Risks: The Culture of Control in the Nuclear Power Industry,* Princeton: Princeton University Press.

Perrow, C. 1999. *Normal Accidents: Living with High-Risk Technologies,* Princetown: Princetown University Press.

Phimister, J.R., Okte, U., Kleindorfer, P.R. and Kunreuther, H. 2003. Near-Miss Incident Management in the Chemical Process Industry. *Risk Analysis,* 23(3), 445–459.

Power, M. 2004. *The Risk Management of Everything: Rethinking the Politics of Uncertainty.* London: Demos.

Rasmussen, J. 1982. Human Errors: A Taxonomy for Describing Human Malfunction in Industrial Installations. *Journal of Occupational Accidents,* 4311–333.

Reason, J. 1990. *Human Error,* Cambridge: Cambridge University Press.

Reason, J. 1997. *Managing the Risks of Organizational Accidents,* Aldershot: Ashgate.

Reason, J. 2008. *The Human Contribution: Unsafe Acts, Accidents and Heroic Recoveries,* Farnham: Ashgate.

Reed, M. 1991. Organizations and Rationality: The Odd Couple? *Journal of Management Studies,* 28(5), 559–567.

Roberts, K.H. 1990. Some Characteristics of One Type of High Reliability Organization. *Organization Science,* 1(2), 160–176.

Roberts, K.H., Stout, S.K. and Halpern, J.J. 1994. Decision Dynamics in Two High Reliability Military Organizations. *Management Science,* 40(5), 614–624.

Robertson, D., Vaughan, J. and Stewart, F. 2008. Strike Force: How the CFS Conquered the Beast. *The Advertiser,* 15 March 2008, p.1.

Rochlin, G. I. 1993. Defining 'High Reliability' Organizations in Practice: A Taxonomic Prologue, in *New Challenges to Understanding Organizations,* edited by K. H. Roberts. New York: Macmillan.

Roe, E. and Schulman, P.R. 2008. *High Reliability Management: Operating on the Edge,* Stanford: Stanford University Press.

Rooksby, J., Gerry, R.M. and Smith, A.F. 2007. Incident Reporting Schemes and the Need for a Good Story. *International Journal of Medical Informatics,* 76(S), S205–S211.

Salas, E. and Klein, G. (eds.) 2001. *Linking Expertise and Naturalistic Decision Making,* Mahwah, New Jersey: Lawrence Erlbaum Associates.

Sandman, P. 1993. *Responding to Community Outrage: Strategies for Effective Risk Communication,* Fairfax, VA: American Industrial Hygiene Association.

Schank, R.C. 1990. *Tell Me a Story: A New Look at Real and Artificial Memory,* New York: Maxwell Macmillan International.

Schein, E. 1992. *Organizational Culture and Leadership,* San Francisco: Jossey-Bass.

Schram, S.F. and Caterino, B. (eds.) 2006. *Making Political Science Matter,* New York: New York University Press.

Schulman, P.R. 1993. The Negotiated Order of Organizational Rationality. *Administration and Society,* 25(3), 353–372.

Schulman, P.R. 1996. Heroes, Organizations and High Reliability. *Journal of Contingencies and Crisis Management,* 4(2), 72–82.

Sills, D.L., Wolf, C.P. and Shelanski, V.B. 1982. *Accident at Three Mile Island: The Human Dimensions,* Boulder, Colorado: Westview Press.

Silverman, D. 2001. *Interpreting Qualitative Data: Methods for Analysing Text, Talk and Interaction,* London: Sage.

Simon, H.A. 1956. Rational Choice and the Structure of the Environment. *Psychological Review,* 63, 129–138.

Sitkin, S. 1992. Learning Through Failure: The Strategy of Small Losses, in *Research in Organizational Behavior,* edited by R.I. Sutton & B.M. Staw. Stamford: JAI Press Inc.

Snook, S.A. 2000. *Friendly Fire: The Accidental Shootdown of US Black Hawks over Northern Iraq,* Princeton: Princeton University Press.

Standards Australia 2001. Occupational Health and Safety Management Systems – Specification with guidance for use: AS/NZS 4801:2001.

Standards Australia 2008. Quality management systems – Requirements AS/NZS ISO 9001:2008.

Sullivan, W.M. 2005. *Work and Integrity: The Crisis and Promise of Professionalism in America,* San Francisco: Jossey-Bass.

Tetlock, P.E. 1992. The Impact of Accountability on Judgement and Choice: Toward a Social Contingency Model, in *Advances in Experimental Social Psychology*, edited by M. P. Zanna. New York: Academic Press, 331–376.

Turner, B.A. 1990. The Rise of Organizational Symbolism, in *The Theory and Philosophy of Organizations*, edited by J. Hassard & D. Pym. London: Routledge, 83–96.

Turner, B.A. and Pidgeon, N.F. 1997. *Man-made Disasters,* Oxford: Butterworth.

Vaughan, D. 1996. *The Challenger Launch Decision: Risky Technology, Culture and Deviance at NASA,* Chicago: University of Chicago Press.

Weick, K.E. 1993. The Collapse of Sensemaking in Organizations: The Mann Gulch Disaster. *Administrative Science Quarterly,* 38(4), 628–652.

Weick, K.E. 1995. *Sensemaking in Organizations,* Thousand Oaks: Sage Publications.

Weick, K.E. 1998. Foresights of Failure: An Appreciation of Barry Turner. *Journal of Contingencies and Crisis Management,* 6(2), 72–75.

Weick, K.E. (ed.) 2001. *Making Sense of the Organization,* Oxford: Blackwell Business.

Weick, K.E. 2006. Faith, Evidence, and Action: Better Guesses in an Unknowable World. *Organization Studies,* 27(11), 1723–1736.

Weick, K.E. 2007. The Generative Properties of Richness. *Academy of Management Journal,* 50(1), 14–19.

Weick, K.E. and Roberts, K.H. 1993. Collective Mind in Organizations: Heedful Interrelating on Flight Decks. *Administrative Science Quarterly,* 38(3), 357–381.

Weick, K.E. and Sutcliffe, K.M. 2001. *Managing the Unexpected: Assuring High Performance in an Age of Complexity,* San Francisco: Jossey-Bass.

Weick, K.E., Sutcliffe, K.M. and Obstfeld, D. 1999. Organizing for High Reliability: Processes of Collective Mindfulness, in *Research in Organizational Behavior*, edited by R.I. Sutton & B.M. Staw. Stamford: JAI Press Inc.

Weick, K.E., Sutcliffe, K.M. and Obstfeld, D. 2005. Organizing and the Process of Sensemaking. *Organization Science,* 16(4), 409–421.

Wenger, E. 1998. *Communities of Practice: Learning, Meaning, and Identity,* Cambridge: Cambridge University Press.

Westrum, R. 1994. Thinking by Groups, Organizations and Networks: A Sociologist's View of Science and Technology, in *The Social Psychology of Science*, edited by W.R. Shadish & S. Fuller. New York: The Guilford Press.

Westrum, R. and Adamski, A.J. 1999. Organizational Factors associated with Safety and Mission Success in Aviation Environments, in *Handbook of Aviation Human Factors*, edited by D.J. Garland, J.A. Wise & V.D. Hopkin. Mahwah, New Jersey: Lawrence Erlbaum Associates.

Wildavsky, A. 1988. *Searching for Safety*, New Brunswick: Transaction Books.

Woods, D.D. 2006. Essential Characteristics of Resilience, in *Resilience Engineering: Concepts and Precepts*, edited by E. Hollnagel, D.D. Woods & N. Leveson. Aldershot: Ashgate.

World Association of Nuclear Operators. 2002. *Principles for Effective Operational Decision Making.*

Wynne, B. 1988. Unruly Technology: Practical Rules, Impractical Discourses and Public Understanding. *Social Studies of Science,* 18(1), 147–167.

Zabusky, S. and Barley, S. 1996. Redefining Success: Ethnographic Observations in the Careers of Technicians., in *Broken Ladders: Managerial Careers in the New Economy*, edited by P. Osterman. New York: Oxford University Press.

Index